M000307849

SPEED READ
CAR DESIGN

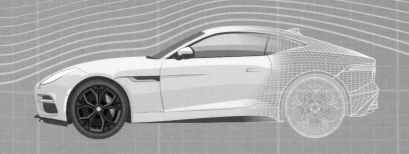

Inspiring | Educating | Creating | Entertaining

Brimming with creative inspiration, how-to projects, and useful information to enrich your everyday life, Quarto Knows is a favorite destination for those pursuing their interests and passions. Visit our site and dig deeper with our books into your area of interest: Quarto Creates, Quarto Cooks, Quarto Homes, Quarto Lives, Quarto Drives, Quarto Explores, Quarto Gifts, or Quarto Kids.

10 9 8 7 6 5 4 3 2 1

ISBN: 978-0-7603-5810-8

Library of Congress Control Number: 2017946394

Acquiring Editor: Zack Miller
Project Manager: Jordan Wiklund
Art Director: James Kegley
Cover and page design: Laura Drew

Cover illustration by Jeremy Kramer
Interior illustrations by Jeremy Kramer except where noted below.
© Shutterstock: p.10-11: *Roman Ya*, p.26-27 & 66-67: *Greens87*, p.48-49: *Denys Po*, p. 94-95: *Gercen*, p.116-117: *Makadoek*, p.132-133: *adike*

Printed in China

MIX
Paper from
responsible sources
FSC® C101537

SPEED READ

CAR DESIGN

THE HISTORY, PRINCIPLES AND CONCEPTS
BEHIND MODERN CAR DESIGN

TONY LEWIN
FOREWORD BY GERRY MCGOVERN

FOREWORD
WHY GOOD DESIGN RESONATES

From a very early age I was inspired by modernism and clean, reductive design. Whether it's the design of a building, a watch, or an everyday object, I truly believe design has the power to enrich people's lives and raise their spirits. The most compelling designs, of course, elevate themselves above the ordinary and create an emotional response.

For me, a product that resonates on an emotional level requires three key ingredients to be truly compelling. The first is visceral: When I look at it, do I desire it? The second is behavioral: When I use it, does it function properly? And the third is reflective: Having experienced it over a period of time, do I continue to desire it, does it still work, and have I built a positive relationship with it?

In my view, when it comes to vehicle design, this emotional connection can be even stronger. A compelling vehicle design demands your attention and can make the difference between success and failure in the marketplace.

Clearly there is a level of subjectivity when it comes to a customer's appreciation of design. However, a professional designer knows that the core of a compelling design is getting the fundamentals right—such as volumes, proportions, stance, how the car sits on its wheels, its overall length relative to its height and width, its wheel-to-body relationship . . . Car design is complex, but if you get these things right, you're well on your way to a successful design.

As an example of this approach, the Range Rover Velar has exquisite volumes and proportions, which combine to create a stunning silhouette. The relationship between its large, 22-inch wheels and its beautifully sculpted wheel arches communicates a finely tuned combination of drama and agility. The interior is a calm sanctuary where technology is hidden until lit, touch-screen smooth, and at your fingertips when you need it. Its design was driven by a single-minded modernist approach throughout.

Design leadership is crucially important for successful vehicle design. It's about taking people with you. If you can take engineers with you and help them, show that you want to support them, and create things that are truly special, you're going to get a much better relationship that's conducive toward creative harmony. The relationship between design and engineering is key.

Communication also plays an important role. Design leaders need to be able to articulate the design vision, while being competent at intellectualizing various arguments to successfully support the design vision—and of course the brand's DNA has to be at one with that vision. Design is the conduit that communicates what your brand represents.

At Land Rover I've tried to embed a vision for the brand that has design excellence at its very core. While accepting that vehicle design is a collaborative and a multi-disciplined activity, there still has to be a singular view at design leadership level because design by committee is a recipe for mediocrity—and I'm not interested in that.

—Gerry McGovern
Chief Design Officer, Land Rover
May 2017

INTRODUCTION

From Model T to Model S, and from rich man's amusement to universal mobility tool, the motor car has come a long way in the 130 years since Karl Benz's flimsy Patent Motorwagen first spluttered into life.

There have been countless small steps along the way, as well as dozens of big strides—some of the most notable being the Ford Model T, which launched mass production to put America and, indirectly, the world, on wheels, and the remarkable Citroen DS and the leap into the future that it provided in 1955. Post DS, the Austin Mini revolutionized the way small (and later, almost all) cars were designed. In 1997, the Toyota Prius introduced the world to hybrid power, followed by Nissan and Tesla, which opened up pure-electric travel.

All these designs were landmarks in improving how a vehicle functions, what it sells for, and how it minimizes its impact on the global environment. Others have enriched the automotive universe by pushing the bounds of beauty, engineering endeavour, or sheer enjoyment—and for this we celebrate them.

In *Speed Read: Car Design* we trace the inspirations of the first car designers and track the craft, the art, and the science that have propelled successive generations of designers and shaped the contours of the vehicles we see all around us. We show how technical advances and tightening regulations have kept designers on the front foot to provide the aesthetically appealing vehicles demanded by consumers and their changing tastes. Finally, we chart the sometimes dramatic shifts in four-wheeled fashion that have provided breakthrough opportunities for some car manufacturers but spelled decline and despair for others.

We'll also catalog the shifting styles that have given each era of automobiles its own distinct personality: classical elegance in the 1920s; fins and excess in the 1950s; cautious mediocrity in the 1980s and 1990s; and the resurgence of individuality and identity that pushed premium brands into the 21st century. Today, there is a sense of being at a crossroads, an inflection point where every player wants to be premium, but with some seeking advantage by differentiating themselves as premium with a clean, green conscience. Much is up in the air right now as the advent of electric power and automated travel makes the accumulated experience of traditional automakers less relevant to the upcoming wave of car buyers. New entrants—such as Tesla—have been gifted a once-in-a-lifetime opportunity.

The challenge now for designers is to motivate a young and urban generation, people for whom a car is a choice, not an obligation. They may avoid vehicle ownership with car-share schemes; they're likely to be more spontaneous and less respectful of past pecking orders; and they are probably swung more by style than what's under the skin. And they won't necessarily be bothered with the conventional wisdoms of power and performance and the need for the driver to be the center of the show.

For this new connected—yet also disconnected—generation, designers will have to find new ways of stirring up emotions with the products we choose to call automobiles—even though they may lack the cylinders, the turbos, and the howling exhausts their parents and grandparents used to dote on. The future in car design will be a challenge indeed.

—Tony Lewin

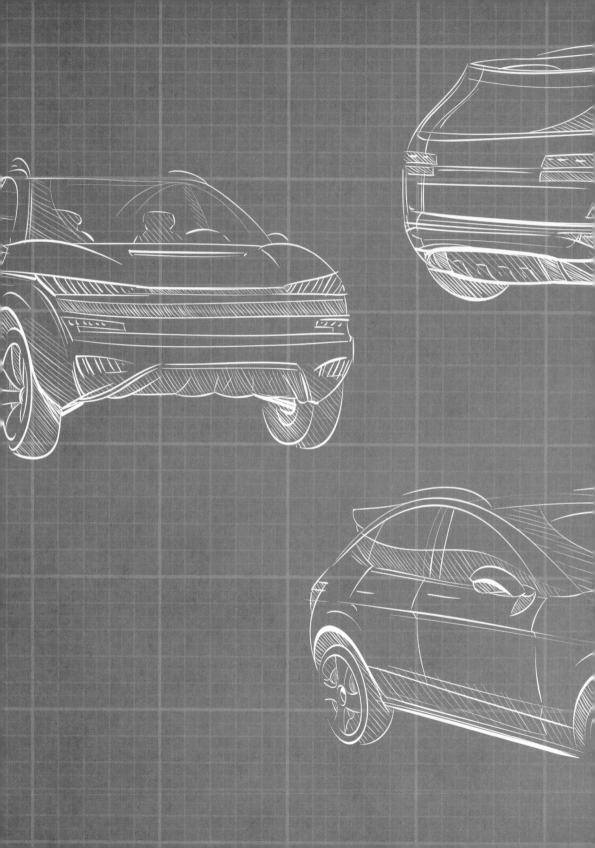

THE BIRTH OF CAR DESIGN

INTRODUCTION
DESIGN—THE GREAT DIFFERENTIATOR

Design had little or no part to play when the very first automobiles sputtered to life. Even engineering—if it could be flattered with such a term—was in short supply as pioneer inventors improvized with available materials from the horse and cart era. Comfort and style were irrelevant: the priority that overwhelmed everything else was that the contraption would start and would not break down too frequently.

Fast forward twelve decades to today's mass-produced models and the position could not be more different. Today, performance and complete reliability are taken for granted, and it is design that sets one model apart from another. The image the car projects—what it looks like, how it feels, and how its quality stacks up—all have become at least as important as its engineering and its responses on the road. In short, design has become the greatest differentiator and style is the new entry ticket to customers' hearts. Cars, like clothing brands, have become industrialized, commoditized, and commercialized. They are close to becoming fashion impulses rather than measured technical choices.

All this has elevated car design—and by consequence, the car designer—to new heights of importance. Now both the architect and the guardian of the brand, the designer is the figurehead giving visual expression to its messages and its values.

But it was not always so. While some of the early proprietors cared little about the appearance of their products, others, especially as the 1920s morphed into the 1930s, became fierce in the protection of the visual message delivered by their models.

For every Henry Ford decreeing that all cars must be black and basic there was a Vincenzo Lancia or André Citroen blending engineering innovation with elegant style; while Charles Rolls and Henry Royce specified their palatial radiator grilles but left the aesthetic shaping of the bodywork to outside craftsmen, there was Ettore Bugatti who, in collaboration with his son Jean, produced complete machines of such exquisite beauty that they are still feted as works of art. Later, William Lyons's intuitive instinct for style ensured that all Jaguars would share the same sense of lithe, sinuous elegance, and in the modern era companies such as Audi have built their reputation on meticulous attention to every aspect of design.

Though Jaguar's Lyons era is separated from Audi by several decades, the two share the common thread of a powerful individual with a keen eye for style and a strong brand image. Today, corporate consensus may have replaced the proprietor's decree, but the message is the same: quality design sells, and is essential to the building of a strong brand position. What is different now is that good design has become democratized: thanks to companies such as Renault, which was a pioneer in bringing in design at a strategic level, true style is accessible to all—not just the fortunate few.

THE BIRTH OF CAR DESIGN
EVOLUTION OF THE SPECIES

Crossovers, shooting brakes, four-door coupés, off-road convertibles: the list of body styles is endless and the variety of formats offered today is truly bewildering. But has this always been the case?

In the 1950s and 1960s, when sedans and station wagons ruled the roost, there was much less variety. This was for sound technical reasons as automakers were just getting to grips with building cars with unitary construction techniques that, initially, made it uneconomic to add further body styles. In the interwar period the selection of styles had been broader: when vehicles were built on a separate chassis, the upper body was not a structural element and it was easier to provide a range of options.

It is possible to trace the origins of almost any body style back to the formats that first began to develop in the early 1900s. Coachbuilders applied their horse-and-carriage styles to motorized chassis: saloons had full roof structures to keep passengers warm, while limousines and broughams left the chauffeur out in the cold. Landaus and phaetons could be opened up at the rear; coupés were simply cut-down versions. Sports cars evolved along a separate path: minimalist and much lower, they remained stark and functional until the aerodynamic 1936 BMW 328.

Greater orthodoxy ruled the postwar decades, with the two-seater sports car a highlight. The 1960s saw the rise of small cars as well as the first examples of a body type that would become dominant—the hatchback. In 1970 the Range Rover set the seeds for the sports utility craze of the late 1990s, which in turn led to the post-2010 crossover phenomenon as automakers merged SUV design cues with mass-production car platforms.

And as the 2020s draw near, further trends are becoming clear: electric cars of all shapes and sizes, the emergence of compact but tough-looking "urban" crossovers, and the polarization of the large-car sector into low-slung, sporty, four-door coupés and more versatile models incorporating design cues from crossovers. And then there is the expected arrival of dedicated zero-emission vehicles locked in to urban mobility programs—and that will be yet another new direction in design.

THE BIRTH OF CAR DESIGN
THE GOLDEN AGE OF CAR DESIGN

BUGATTI TYPE 41 ROYALE
Close to seven meters in length, the Bugatti Royale was one of the grandest cars ever built. Only six were made.

POSTWAR: BUSINESS NOT AS USUAL
Following World War II there was little appetite for the extravagant coach-built creations of the prewar era and many of the big names went under. American cars continued to get longer and larger, but in Europe only Facel Vega, Ferrari, and Maserati retained real glamour.

RETURN OF OPULENCE
Early 21st–century opulence was most likely to be found in an oversized SUV such as the militaristic Hummer or the giant Cadillac Escalade. But Rolls-Royce's revival under BMW ownership prompted a parallel renewal of VW-controlled Bentley and, less successfully, Mercedes's bid to restore Maybach with a stretched S-Class.

If ever there was something that could be called a golden age for sheer style, it must be the 1920s and 1930s—the Jazz Age. For cars, it had all the right ingredients: they were the newest and most exciting phenomenon of the era, technology was advancing rapidly, and there were no tiresome rules and regulations to limit the creativity of designers or the fantasies of customers.

And it was the customers, invariably rich and out to make an impression, who were the key. This was the age that took in the rise of wealthy industrialists and highly paid Hollywood movie stars; at the same time, the world's aristocracy still had cash to burn and viewed the commissioning of a bespoke, one-of-a-kind car in much the same way as the acquisition of a work of art. Indeed, that is what many of the extravagant designs of this era effectively were—extravagant rolling sculptures that served to proclaim the owner's exquisite taste wherever he or she chose to drive.

The prevailing industry culture helped too. Coachbuilders such as Barker and Mulliner in the UK or Chapron in France would shape and craft a unique body style onto a chassis chosen by the customer, giving them almost unlimited freedom for imagination, artistic expression, and the use of sumptuous materials. All Rolls-Royces were produced in this way.

Standout names in this most glamorous of eras include Lincoln, Cord, Duesenberg, and Auburn in the US; Lagonda and Daimler in Britain; and Voisin, Delahaye, Delage, and Talbot Lago in France; as well as Mercedes and Benz—separate firms until 1926—and Horch in Germany. Among coachbuilders, art deco masterpieces from France's Figoni et Falaschi and Saoutchik count among the most imaginative, with huge, sweeping lines, boat-like tails, and exaggerated front fender fairings.

In parallel, a technology race was taking place, with almost all top-end producers including a V–12 engine in their offerings. Cadillac even went to a V–16, though for the ultimate in opulence the 1926 Bugatti Royale was never matched. Sadly, the age of extravagance was cut short by war.

THE BIRTH OF CAR DESIGN
POWER TO THE PEOPLE

FORD MODEL T
Light, simple, and rugged, the 1908 Model T took 93 minutes to build, compared to 12 hours for its predecessor. With a price that dropped below $260 it could be afforded by everyone, even the workers who built it, triggering a social revolution.

TRABANT: THE IRON CURTAIN ANSWER
No objective account of cars for the people would be complete without mention of the Trabant. This tiny, plastic-bodied two-door with its sputtering two-stroke engine mobilized most of eastern Europe in the 30 years leading up to the fall of the Berlin Wall in 1989.

TATA NANO
Hailed by CEO Ratan Tata as the car that would get millions of motorcycle-riding Indian families into much safer cars, the 2008 Nano targeted a price of $2,000. But the figure soon crept up, buyers rejected the flimsy build and egg-like shape, and the model flopped.

Mobility for the masses had been the dream of engineers from early Victorian times, but it only became feasible when a certain Henry Ford rethought the manufacturing process to bring his new Model T onto the market in 1908 at just 850 dollars and later dropped the price to $260.

The massive success of the Model T put America on wheels and is the stuff of history. Europeans soon copied Ford: by 191, Morris's low-cost Oxford was in volume production and Herbert Austin went a step further and designed the tiny Austin 7, the Baby Austin. This was a big seller in many markets and even formed the basis for BMW's first car in Germany.

In Italy the compact Fiat 500 Topolino, designed by the legendary Dante Giacosa in 1936, was instrumental in mobilizing much of the nation; it only ceased production in 1955—to be replaced by the larger, rear-engined 600. Giacosa followed with the Nuova 500 in 1957, another remarkable success. The even more basic Citroen 2CV, first aired in 1938, paralleled the Fiat and lasted until 1988, while Volkswagen's Type 1 Beetle, the self-proclaimed people's car, survived through to 2003.

Fuel and material shortages across Europe in the 1950s had led to the rise of tiny bubble cars, and BMW cashed in on the boom with its two-seater Isetta. In Britain, BMC (the fusion of Austin and Morris) responded in 1959 with the Mini, a wholesale rethinking of small-car architecture that placed the engine across the front of the car for compactness and agile handling. But while the Mini was great fun, it was barely profitable and astute operators such as Ford realized more money could be made with simpler but larger cars.

Before long, mini-sized cars had become sophisticated superminis such as the Renault 5, Peugeot 205, and Ford Fiesta; growing concerns about safety had made it difficult to build truly small cars more cheaply than compact ones, and larger models became the focus. It was only in 2000, with the Smart, that small-car talk temporarily returned. Now, with mature markets stagnating and developing regions offering the main opportunity for expansion, the big sellers are again likely to be larger but simpler, along the lines of the Renault Logan.

HARLEY EARL

Harley J. Earl may not have been the world's first professional automobile designer, but he was certainly the first to achieve celebrity status. He was able to do this because of his intuitive understanding of what motivated people to choose a particular car. Crucially, he was the first to realize that for the vast mass of buyers a car was not so much about technical engineering or performance, but more about the vision of the future that it presented.

Earl was born into the family coachbuilding business in Los Angeles, and it was a natural step for the firm to begin customizing vehicles for the rising generation of wealthy stars and entrepreneurs that was springing up around the nascent Hollywood movie industry. This gave Earl a flair for design, and soon his talents were spotted by Alfred P. Sloan, president of General Motors—which at that time comprised Buick, Chevrolet, Cadillac, Oldsmobile, and Oakland, which would later become Pontiac.

Sloan was an individual of immense significance to the auto industry. He was the first to come up with a modern market segmentation approach—a hierarchy of brands to suit buyers of "every purse and purpose," with Chevrolet at its base and Cadillac at the apex. Sloan quickly saw how someone with a talent—such as Earl's—for turning dreams into metal could make this brand strategy more powerful still: design and styling would be integral parts of marketing rather than afterthoughts of the engineering function.

Thus it was that in 1927 GM established its Art and Colour Section, the first self-contained design studio among automotive manufacturers. Earl introduced the idea of modelling clay to demonstrate the studio's ideas in three dimensions: an early outcome was the Cadillac LaSalle, and important initiatives included the development of a palette of colors and trims—in stark contrast to Henry Ford's grim black-only regime.

New designs soon followed, establishing a pattern of annual styling revisions to keep customer interest at a high level. The 1933 Chicago Word's Fair was the perfect opportunity for Earl to showcase the latest and most futuristic innovations under development at GM: the Cadillac V16 Aero-Dynamic Coupé pioneered all-steel construction, especially the one-piece roof structure, which was later marketed as the "turret top."

By 1938 Earl's vision had become more ambitious still, as demonstrated by the Buick Y-Job, widely seen as the world's first concept car in the sense that it was not intended for production, but to test public reaction to new and advanced design and engineering ideas. Many of these ideas, such as the electrically raised and lowered roof, would appear in the late 1940s as another of Earl's obsessions—jet aircraft—began to influence production models. Cadillac was the first to sport fins in 1948, a craze that was to last more than a decade. Other even more futuristic "research" vehicles would appear throughout the 1950s as GM's Motorama roadshow toured the US, casting over a whole generation of car buyers an excited spell from which they would never recover.

THE BIRTH OF CAR DESIGN
THE EMERGENCE OF STYLE

FAST FASHION
Updating a car's styling every year took hold in the US in the 1930s as automakers followed GM's example to compete for fashion-conscious buyers. The practice developed into near hysteria in the 1950s but had calmed down by the 1970s, when safety and emissions became greater concerns.

ANTI-STYLE STATEMENT
The central theme of Volkswagen's marketing in the US in the 1950s and 1960s was to mock the lack of style of its own Beetle/Bug, mischievously drawing attention to the excesses and planned obsolescence of its American competitors.

STYLE-LED BRANDS
Continuity of style or image is easier to achieve when one man or one family is in charge. William Lyons's intuitive design flair ensured grace and feline elegance for all Jaguars up to the late 1960s, and Porsche has kept strict control on the lines of all its sports cars since the original in 1948.

When the Ford Model T launched the era of mass production in 1908, the motorcar became something almost every working person could aspire to. In Europe, where Ford's moving assembly lines were widely copied, some firms concentrated on further improving efficiency and reducing prices, while others sought to improve their designs by introducing new engineering features.

In the US there were also some who thought differently. Ford's cost-cutting approach had been astonishingly successful, but others saw this as a race to the bottom: with the vehicle remaining simple and durable and barely changing over the course of many years, owners had no reason to upgrade to a new example to keep the factories running full tilt.

By contrast, Ford's rival General Motors launched itself into a sophisticated marketing strategy involving a hierarchy of brands and employing innovative styling rather than engineering or cost-cutting as the main marketing tool. Car design had come of age, and it was a formula that worked so well that competitors were forced to follow suit with design operations and star designers of their own. Cars had now become items of fashion rather than mere assemblages of engineering features, and the annual model styling refreshes were eagerly awaited.

The styling arms race reached its zenith in the 1950s as designers rushed to reshape every model every year, leading to enormous excesses in ornamentation, elaboration, and sheer size. Some of these traits crossed the Atlantic to appear on US-owned European brands like Ford, Taunus, Opel, and Vauxhall, contrasting with traditional automakers such as Peugeot and BMC, which had developed their own distinct design identities through collaboration with design houses such as Pininfarina.

But for every company espousing the value of elegant design there were still several that insisted that engineering should take priority. Roll forward to the 21st century, however, and those voices are silent: today, with performance and complete reliability taken for granted, it is design that sets one model apart from another and the designer, not the engineer, is the key to success.

GLOSSARY: CAR BODY STYLES AND TYPES

Berline (France), Berlina (Italy): Sedan with separate trunk.

Berlinetta (Italy): Sports coupé.

Break (France): Station wagon.

Boat-tail: Long, tapering tail reminiscent of a boat's; popular in the interwar period and revived by Alfa Romeo on the 1960s "graduate" Spyder.

Brougham (US): Classic carriage style with open front and closed rear.

Cabriolet, cabrio: Version of sedan or hatchback with opening roof.

Convertible: Vehicle with generally full-length opening roof.

Coupé (France): Literally, "cut." Sports model with lower roofline, lower build, and reduced (or no) rear seat accommodation.

Crossover: Vehicle type with higher silhouette but not necessarily four-wheel drive or off-road capable.

Crossover coupé: SUV or 4x4 with fastback rear roofline and/or lower silhouette.

Drophead: Classic term for folding roof version of fixed head [roof] coupé.

Estate car (Great Britain): Station wagon.

Fastback: Sloping rear roofline to create a two-box profile, as opposed to the three-box profile of a sedan.

4x4: General term for off-road-capable vehicles with all-wheel drive.

Hardtop: Originally, a coupé or sedan without the center (B) pillar.

Hatchback: Body style where the trunk lid hinges above the rear window and the rear seats can be folded for load carrying.

Kombi (Deutschland): Station wagon.

Landau, Landaulet: Classic configuration with enclosed front compartment and openable rear section. Often used for dignitaries on state occasions.

Limousine: Large car, generally chauffeur-driven, with partition separating driver from passengers. Now applied to stretched versions of standard luxury cars.

Minivan (US): Versatile vehicle with taller silhouette carrying seven or more passengers. Also known as MPV or people carrier.

Phaeton: Classically, a large, fully open luxury car. Used by Volkswagen as a model name.

Roadster: Two-seater open sports car, sometimes without full windshield

Sedan (US), Saloon (United Kingdom): Three-box design with separate trunk. Contrast with hatchback, station wagon.

Sedanca: Vehicle with open driver's compartment but closed accommodation for passengers in the rear.

Shooting Brake: Originally, station wagon with lower roofline. Now applied to sportier station wagons with fastback rear profile.

Sports Utility Vehicle (SUV): General term for high-riding wagon-style vehicles suggesting off-road ability; not necessarily 4x4.

Spyder, Spider (Italy): Open-topped sports car or coupé derivative.

Station Wagon: Standard car with extended roofline, rear tailgate, and folding rear seats.

Targa: Originally, Porsche term for car with removable roof panel(s) leaving the rear window in place.

Torpedo: Classical sports car body style with cut-down door openings.

Tourer: Originally, a convertible version of a sedan. Now a name for station wagons.

Town Car: Classical body style with open front and closed rear.

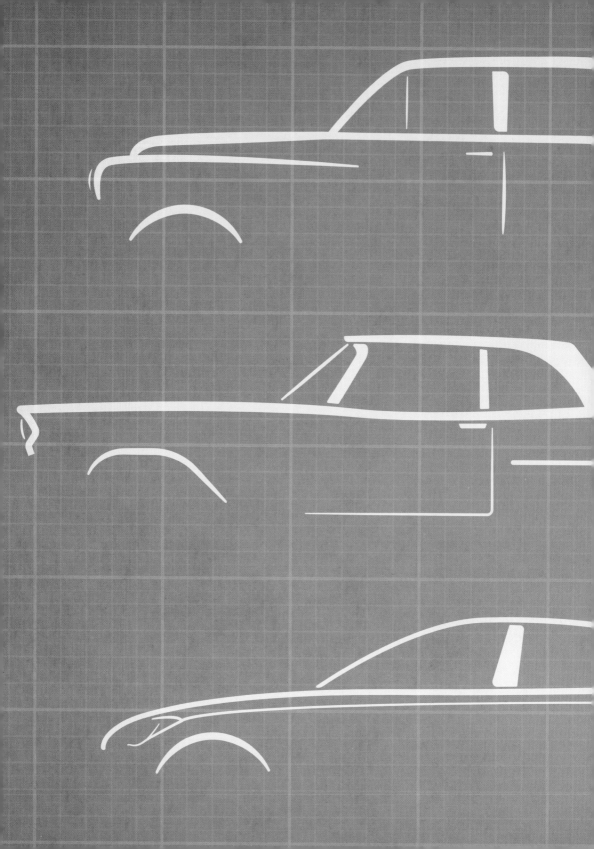

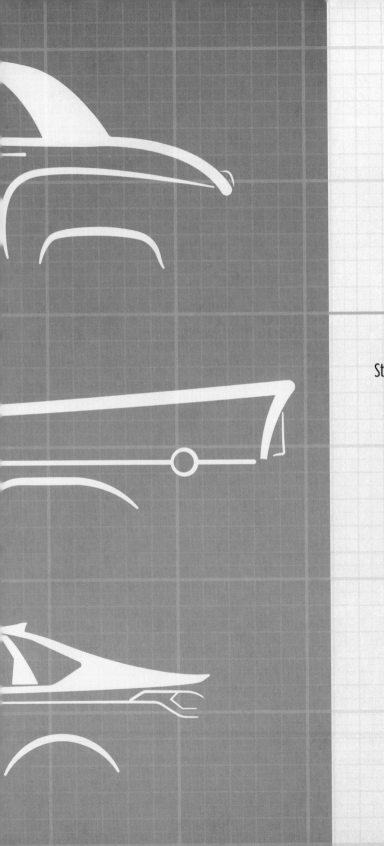

A CENTURY
OF CHANGING
SHAPES

INTRODUCTION
THEMES, STYLES AND TRENDS

Just as there are different movements and fashions in architecture, so is the world of car design steered by social trends and shifts in consumer taste. And as with the evolution of building styles, the shapes and characters of cars have progressed as new materials and manufacturing techniques are developed. Smooth, all-enveloping plastic front ends are a case in point: over the course of the 1970s and 1980s they replaced the traditional chrome metal bumpers, transforming the look and feel of cars in the process.

Technology has driven many of the major shifts in vehicle design over the years. Wider steel-rolling mills in the 1930s allowed larger pressings, in turn helping cars move from incoherent assemblages of separate elements such as fenders, hoods, and running boards to the smooth-sided "pontoon" look that dominated the 1950s; new steel blends and clever pressing techniques allowed designers to specify sharper creases and more complex shapes toward the end of the century.

Such developments have helped the outline of the car to branch out from the classical three-box format to smooth two- and one-box shapes. At the same time, the basic proportions in terms of height, width, and glass-to-metal ratio have fluctuated dramatically. Likewise, evolution in mechanical layouts has led to big changes in the positioning of the passenger cabin in relation to the hood, wheels, and trunk: witness Chrysler's "cab forward" architecture of the 1990s or the exaggerated rearward cabin bias of the BMW 1-Series in 2004.

Leading designers over the years have been quick to exploit engineering developments to launch new styles and fresh ways of linking surfaces. Consider Giorgio Giugiaro's 1974 Volkswagen Golf, which introduced a crisp and angular surface language that was widely copied. The Golf's hood line was new too, sloping downward as it ran forward from the windshield for a wedge profile that also became near universal.

Often it is the surface language of a design—the way one surface transitions into its neighbor—that is the most characteristic of a style or an era. Smooth, large-radius transitions were typical of the 1980s "aerodynamic" age, while sharper definitions and tighter radii gave more punch to later forms. The early 2000s in turn witnessed a profusion of crease lines and surface changes within panels, especially body sides, while a recent trend in some designs for greater overall simplicity is achieved through complex surface profiles in key areas such as the hood-fender transition and the base of the windshield. The Audi Q8 is just such an example.

In parallel, a myriad of developments in other areas such as paint, color, and lighting is allowing designers to transform the visual signatures of their vehicles. Fiber optic pinpoints and LED strips, for instance, provide the opportunity to dramatically change graphics, even in real time, and technology is under development to display images, textures, and different colors on body panels. Panels can even change shape too. The scope is immense: rarely have designers had so many elements to play with, and never has the future of car design been more exciting.

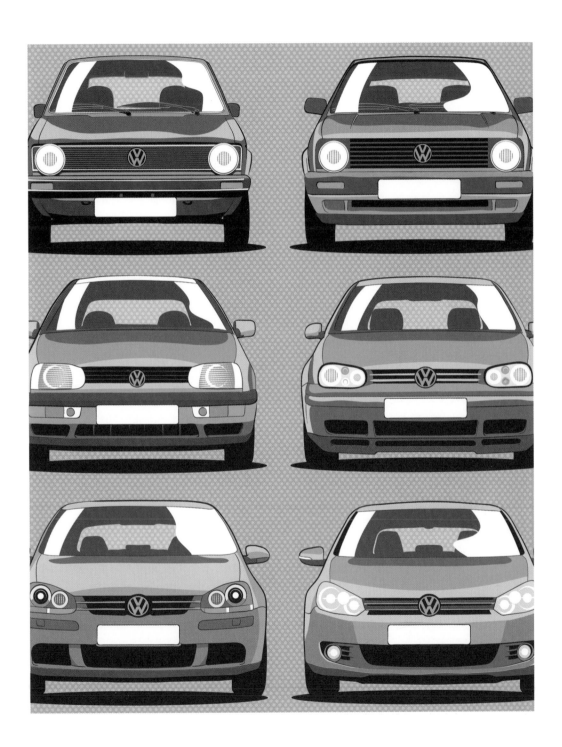

A CENTURY OF CHANGING SHAPES
STREAMLINING: A VISION OF THE FUTURE

DRAGGING THE LINE
A car's Cd, or drag factor, is a measure of the slipperiness of its shape. The lower the figure the better, and today's best cars are between 0.30 and 0.21. The Cd value is multiplied by the vehicle's frontal area to calculate actual wind resistance.

THE SQUEEZE IS ON
Pininfarina's CNR "Banana Car" concept shocked the automotive establishment in 1978 with its flowing shape squeezed in the center for efficiency. Its Cd was tested at 0.161, vastly better than the 0.40-plus of the era's typical cars; many feared that computer optimization would make all cars look the same.

EASY FLOW, MORE GO
The most aerodynamic production car to date has been Volkswagen's minuscule XL1 two-seater coupé. Aided by its narrowness and long tail, it rates a Cd of 0.186, close to what is theoretically possible. Electric cars, requiring less cooling air, are expected to be tomorrow's best performers.

The aerodynamic form has been an enduring fascination amongst automobile designers from the earliest days. The streamlining obsession truly took hold in the 1920s—initially inspired by giant Zeppelin airships and, later, by fully enclosed record-breaking cars and motorcycles.

While car factories of the era were content to turn out popular, square-rigged models, imaginative designers such as Paul Jaray and Hans Ledwinka of Tatra came up with sleekly streamlined passenger car concepts. These influenced German engineer Ferdinand Porsche and indirectly led to the famous Volkswagen Beetle.

In the US, Chrysler took a brave step forward with its 1934 Airflow sedan, an elegant art-deco style streamliner that was admired by all—except the buying public, for whom it was too futuristic. But with car factories in full postwar swing in the late 1940s, new and more smoothly integrated shapes began to displace the disjointed boxes of the prewar years.

Swedish aircraft maker Saab's 92 marked a streamlining high point in 1949, but for sheer impact nothing could top the revelation of the 1955 Citroen DS. The sleek, smooth, and low-set DS was so stunningly futuristic that, in an instant, every other car was made to look incoherent and old-fashioned, and for a further quarter century the French company led the way in aerodynamics.

The 1961 Jaguar E-Type was brilliantly successful in translating the voluptuous aerodynamic shape of Jaguar's Le Mans racing cars into a road-going sports car, but a key landmark at the end of the decade was NSU's revolutionary Ro80. Elegantly rather than extravagantly visionary, it paved the way for a new generation of streamlined designs in the 1980s—most notably the Audi 100, first to post a drag factor (Cd) below 0.30.

The tightening focus on emissions and fuel consumption since the 1990s has made every new design more aerodynamically efficient than its predecessor. But with computer-aided engineering able to optimize airflow by looking at small details rather than the overall shape, the days of the standout streamliner could now be over—much to the regret of many.

A CENTURY OF CHANGING SHAPES
FORM VERSUS FUNCTION

**JAGUAR—
MASTER AERODYNAMICIST**
The multiple Le Mans–winning Jaguar D-Type, designed by master aerodynamicist Malcolm Sayer, provided a clear visual demonstration of the value of streamlining in sustaining very high speeds over a 24-hour endurance race.

AUDI TT: LATTER-DAY BAUHAUS STYLE
The smooth and rounded geometric forms of the petite Audi TT coupe gave it instant appeal not just to car fans but a much wider circle of industrial and product designers who valued its Bauhaus-style purity. It added drama to Audi's worthy but then unexciting image.

BATTERY-POWERED CARS DON'T NEED GRILLES
The motors that power electric cars do not need cooling air in the same way as a gasoline or diesel engine does. So, if form is to follow function, why do electric cars still feature visible air intakes at the front? The answer must lie deep in the roots of brand identity.

It is a quandary that has existed ever since people first began making objects. Which is more important—form or function? The rise of manufacturing industries in the 19th century only served to amplify the dilemma.

For the early automobile builders the priority was clear: above all, the machine had to function, and appearance was secondary. Only the wealthy few could afford the stylistic flourishes that came with the extravagant creations from specialist coachbuilders. The most celebrated of these designs were art deco in style, but in Germany a very different movement was taking shape at much the same time. The Bauhaus ethos favored purity and simplification, with form arising from function, not ornamentation or decoration. The early Citroen 2CV is an example. Of today's automakers it is perhaps Audi that best honors those Bauhaus principles.

Those qualities were hard to find in the prewar period, though engineer-led producers such as Bugatti, Citroen, and Lancia achieved greater visual coherence thanks to technical advance rather than deliberate style. It was again left to sports cars to demonstrate their function through their form: low set and streamlined, with long, louvered hoods and smoothly tapered tails, their whole stance spelled speed and power.

In the postwar period, while the US obsession with fins and jet plane ornaments saw form completely overwhelm function, European sports cars such as the Mercedes 300SL, Porsche 356, and Jaguar XK series demonstrated precisely the opposite. BMC's brilliant engineer Alec Issigonis's refusal to allow stylists near his 1959 Mini resulted in a plain but highly effective car that went on to become an icon—a great, if accidental, example of form following function.

More recently, Chrysler's boxy Voyager minivan from 1984 counts as a clear depiction of function, as does Audi's compact and inspirational 1998 TT sports coupé, surely the most eloquent, albeit belated, evocation of the Bauhaus ethos yet seen.

A CENTURY OF CHANGING SHAPES
THE NEED FOR SPEED

HEYDAY OF THE SPORTS CAR

The 1950s and 1960s were the heyday of the two-seater sports car, and US buyers could not get enough of them. Star performers were the MGA and MGB, the Triumph TR series, Chevrolet's Corvette, and the Austin Healey 3000. Alfa Romeo typified the breed with its Duetto Spider, made famous by Dustin Hoffman in *The Graduate*.

TRACK DAY THRILLS

Ever-tighter rules on road driving have prompted thrill-seekers to take to the racetracks. This has spawned a generation of track-day specials such as the Caterham Seven, Ariel Atom, and KTM X-Bow, echoing motorcycle design with their mechanical elements exposed.

LIGHT AND AGILE

Lotus founder Colin Chapman's maxim "just add lightness" resulted in a series of ground-breaking sports cars. Among the all-time greats are the Elan (1962-1973) and the mid-engined Elise (1996-present).

Ever since drivers first took to the road more than a century ago, there have been some who wanted to go faster than others. Some managed this by being braver or more skillful, while others modified their cars for superior performance; it was these vehicles that evolved into racing cars and sports cars—initially there was little difference.

As the exhibitionists of the automobile world, sports cars advertise their speed and power through their design. Early examples favored lower bodywork, with long louvered hoods suggestive of a big, powerful engine, big wire wheels, and more clearly visible brakes—an unashamedly technical approach with the mechanical elements on display rather than hidden away.

That's a design code that holds true to this day, though aerodynamics now has a big part to play and designers have to be more ingenious in how they choose to symbolize power: external exhaust pipes are no longer an option. Mercedes-Benz's 300SL of 1955 was a perfect example: low, smooth, and streamlined, it portrayed power through its numerous air intakes and outlets, speed via its wheel arch eyebrows and subtle use of chrome side strakes, and stability through its wide, planted stance on the road.

Not all sports cars are about sheer power, however, and there is an equally fine tradition of lightweight models typified by today's Mazda MX-5. Many owe their inspiration to Colin Chapman's left-field thinking with the 1957 Lotus Elite, an ultra-light coupé with a tiny engine but remarkable speed and agility. Chapman had already produced the Lotus 7, a very basic design with separate fuselage and fenders: this, too, evolved into a distinct bloodline of minimalist sports cars including today's Ariel Atom and KTM X-Bow.

In the 1970s many feared that US safety regulations would kill off sports cars. Yet in Europe a new trend was about to emerge—the hot hatch. Initially, these were more powerful versions of familiar family hatchbacks such as the Volkswagen Golf, but gradually a new design iconography emerged: wider wheels, lowered suspension, prominent multiple exhaust outlets, and aerodynamic accessories. By the turn of the millennium this had become the dominant sporting genre.

A CENTURY OF CHANGING SHAPES
SMALL IS BEAUTIFUL

THE CLEVER CITY CAR THAT FLOPPED

The Peugeot 1007's tall profile and cumbersome sliding doors restricted its appeal, despite a generous four-seater cabin and excellent versatility within a road footprint of 3.7 meters. It lasted barely four years from its 2005 debut.

THE TINIEST CAR OF ALL

Guinness World Records certifies the 1962 Peel P50 as the world's smallest car. The single-seater three-wheeler measured just 1.3 meters in length and used a 50 cc engine. There was no reverse gear: the driver could simply lift the 56-kg machine and swivel it.

ONE PLUS TWO EQUALS T25

Former racing car designer Gordon Murray has come up with the blueprint for a small car that is cheap as well as light. The T25 sits the driver centrally, with two passengers just behind. Power is either gasoline or electric, but no big manufacturer has yet taken up the idea.

Automotive designers have always taken a certain pride in creating small cars. Engineers relish the technical task of packaging the passengers, the luggage, and all the required mechanical components into a small road footprint, while for the stylist there is the challenge of coming up with a compact shape that looks engaging, credible, and not too toy-like.

Sir Herbert Austin gambled heavily with his new 7 in 1922, a tiny four-seater that was a proper car in miniature: its success killed off the cyclecar industry for good. The same streak of engineering ingenuity led BMC's Sir Alec Issigonis to rethink the rules completely in the 1959 Mini, mounting the engine across the chassis at the front to free up generous accommodation for four in a three-meter footprint. It is a packaging solution that has never been bettered and it totally transformed the auto industry.

However, it is not possible to directly compare Mini-era compact cars with their more modern counterparts for the simple reason that completely different safety standards apply. When the Mini was designed, there were no rules. Now, all cars must protect their occupants when slammed into a barrier at high speed: designers have to allow for crush space front and rear, further increasing the challenge in keeping a small car small.

This calls on even greater inventiveness from the designer in maximizing space efficiency. The 2001 Smart measured just 2.5 meters but seated only two; the current four-seat version is a commendable 3.5 meters, still small by today's standards. The Mitsubishi i also mounted its engine at the rear in a narrow 3.4-meter four-seater package, while the VW Up and Peugeot 108 measure around 3.5. The reborn BMW-style Mini was a relative giant at 3.6 meters, with little more interior space than the 1959 original; its latest incarnation measures over 3.8 and tramples over Issigonis's notions of space efficiency.

But for a modern paragon of clever mini-car packaging, look no further than Toyota. The 2008 iQ cleverly rethought the differential, the heating, and the fuel tank to find space for four—three adults and a child—within the same 3-meter length as the original Mini, but with modern-day safety.

A CENTURY OF CHANGING SHAPES
TOUGH TALK

LAND ROVER SHAPED IN SAND

The design of the original Land Rover was sketched out not on a drawing board but in sand on a beach in Wales, where technical director Maurice Wilks used a Jeep on his farm and saw the opportunity for a Rover equivalent, which became the 1948 Land Rover.

BODY ON FRAME

The traditional structure for a 4x4 was to build a body on top of a strong, separate ladder chassis. But this form can be heavy, crude, and unrefined, and in recent years many designs have moved toward unitary construction, familiar from standard cars, for efficiency.

FROM LIGHTWEIGHT 4X4 TO CROSSOVER

The success of lightweight 4x4s such as the Suzuki SJ and Daihatsu Rocky in the 1980s prompted Toyota to launch the car-based RAV4 in 1994 as a good-to-drive leisure 4x4. Praised as the first GTI off-roader, the RAV helped kickstart the shift to crossovers.

It all began in 1941, with the US Army Jeep. Rover engineers in the UK were inspired to emulate the design, which by then was available to civilian customers too, and came out with the Land Rover in 1948; Toyota followed suit with its Land Cruiser in 1951. All three shared the same strictly utilitarian design traits: simple high-riding angular bodywork with flat metal and glass surfaces, large separate fenders housing chunky tires with plentiful ground clearance, crude interiors, and, of course, four-wheel drive.

This template barely shifted for two or more decades, though in 1962 a new arrival, little noticed at the time, would prompt a massive change by the end of the century. The Jeep Wagoneer, in effect a high-riding station wagon with all-wheel drive, brought the first taste of civilization into 4x4 driving. With the Wagoneer as its inspiration, Land Rover sensed an opportunity in a smarter 4x4 that the farmer or landowner could drive to dinner in the evenings after working on the land during the day. Thus the 1970 Range Rover was born, its classily elegant design (by David Bache) proving an instant hit worldwide and eventually prompting the entirely new genre of vehicles today know as SUVs.

Design cues for SUVs have polarized over the years. US-focused designs tend to be much larger and more truck-like—tall, blocky and intimidating, and with exaggerated grilles, fenders, tires, and running boards. Outside North America there is a gentler feel, with more curves and less aggression; interiors in each case are sumptuous, reflecting luxury price tags.

The big development in recent years has been in the rise of the crossover, an appellation that arose as automakers began to apply SUV design cues such as raised ride height to lighter and more refined passenger car platforms. The inclusion of four-wheel drive was not essential: what mattered to buyers was the tough, go-anywhere image, the elevated driving position, and the versatile passenger and load-carrying arrangements.

In this way, crossovers such as the Nissan Qashqai became a major force in the marketplace, bringing together not only the two strands of SUV culture but also the best aspects of standard family hatchbacks.

A CENTURY OF CHANGING SHAPES
ROOM TO MOVE

THE FIRST PEOPLE CARRIER
The consensus is that this was Chevrolet's 1935 Suburban, an eight-seater wagon built on a half-ton truck chassis. By the eleventh generation in 2015 the Suburban had grown into a 5.7-meter SUV giant with luxurious seating for nine.

WHAT IS THE SMALLEST SEVEN SEATER?
A leading candidate could be the 1999 Daihatsu Atrai, also sold as the Toyota Sparky. These pint-sized Japan-only microvan-based twins managed to cram seven people into a footprint just 3.7 meters long, the tiny engine sitting under the floor.

UGLY DUCKLINGS
Practicality and style do not always go hand in hand. While some prefer the wacky duck-like shape of the Fiat Multipla, the bulbous high-roofed Toyota Yaris Verso and bizarre double-decker Ssangyong Rodius have fewer followers. Even super-successful Skoda polarizes opinions with its Roomster mini-wagon.

For drivers wishing to transport more than four passengers or bulky loads, the choice of vehicle was once seriously restricted—minibus or van, both with a truck-like image and dismal dynamics. The station wagons with three rows of seats that appeared in France and the United States in the 1950s and 1960s were more civilized, but it was not until the arrival of Chrysler's Voyager in 1983 that comfort and space for seven were successfully combined.

That slab-sided box proved highly influential—not in style, perhaps mercifully, but by inspiring a new format for automotive design. The futuristic Renault Espace, with its sleek shape and seven removable seats, led the charge in 1984, making multi-seat family travel swift and exciting. Before long, copycat people carriers had sprung up in most mainstream model ranges, though the Espace continued to define the genre and set the design agenda with its flowing one-box silhouette and commanding driving position.

However, the rise of the tough-looking SUV and sporty crossover in the early 2000s hit the minivan hard. Buyers rebelled against the homely "soccer mom" image of the people carrier and signed up for the more glam-ourous lifestyle of zeitgeist products such as the Volvo XC90 and BMW X5 that combined the advantages of SUVs and station wagons, as well as carrying the full complement of seven passengers. Significantly, Renault re-launched the Espace for its fifth generation in 2015 as a crossover-like vehicle, sportier in allure with a shallower glasshouse and a more expressive body style.

The need to carry five or more people has inspired many novel seat-ing configurations—beginning with the London taxi, which hosts its five passengers in a face-to-face layout. Fiat's original rear-engined Multipla sat six within a length of 3.5 meters, with the driver and front passenger above the front axle. Its brilliant but eccentric namesake of 1998 had a 3+3 arrangement, leaving plenty of luggage space even when all seats were occupied; it spawned an imitator from Honda in 2004, the FR-V.

Strange profiles are another byproduct of the search for space. Japan has a longstanding tradition of box-shaped minivans: the Nissan Cube and Honda Element have become cult cars beloved of design purists.

A CENTURY OF CHANGING SHAPES
RETRO DESIGN

RETRO HORROR STORY

Many ill-conceived models appeared in the rush to cash in on the post-2000 nostalgia craze. Among the more embarrassing are the Chrysler PT Cruiser, Chevrolet HHR rival, and the Plymouth Prowler hot rod.

THE GENUINE ARTICLE?

Could Britain's most unusual car builder, Morgan, also be accused of cynical retro design? Its handcrafted sports cars have always been vintage in their allure, even though the underlying engineering has come up to date, but the eccentric Three-Wheeler of 2011 is an unashamed throwback to the firm's 1930s model.

VW MICROBUS: WAITING FOR A REMAKE

Kindling fond memories of hippy road trips and camping expeditions, Volkswagen has shown no fewer than four studies harking back to the 1950s Microbus: the 2001 Microbus Concept, the 2011 Bulli, and the battery-powered Budd-e and ID Buzz in 2016.

Trying to drive forward while looking in the rear-view mirror is not a good way to make progress—and many believe the same thing applies to the philosophy of retro design. Yet product planners continue to prioritize retro-themed new models for the simple reason that they sell well and their special status makes them formidable profit earners.

Just look at the runaway success of the reborn Mini, Fiat 500, and Ford Mustang: undeterred by inflated prices, excited customers queue up to buy into the period lifestyle, adding costly extras to personalize their purchases. The nostalgia dimension adds authenticity to what would otherwise be an ordinary product. Understandably, the carmakers' finance directors love these high-margin pseudo-premium products—but the designers themselves are not so sure. For many, retro design is an unhealthy trend. Some argue it smacks of a lack of imagination, like a recording artist remixing their greatest hits; others counter that it shows respect for the design back-catalog, honoring a brand's heritage and history.

The first and perhaps most celebrated flowering of retro design came in Japan in the late 1980s when the economy was booming and there was money to burn. Under Naoki Sakai, Nissan produced a remarkable series of four genuinely affectionate retro-themed small vehicles: the Be1 and Pao were Japan-only minicars, while some examples of the exquisite Figaro mini-roadster and circle-themed S-Cargo van did make it to western markets.

Perhaps inspired by Nissan, future Ford design boss J Mays penned the 1994 Concept 1 while still at Volkswagen. This stylized remake of the original Beetle prompted a frenzy of nostalgia, which persuaded VW to put the car into production for the 1997 model year as the New Beetle. It was an unexpected success and, after his move to Ford, Mays launched into full retro mode with the low-rider Forty-Nine in 2001 and the Thunderbird Coupé the following year, promoting the term "retrofuturism" along the way.

The retro craze was short lived, however: the Ford sold poorly, GM's response was half-hearted and only Chrysler's gangster-style 300 survived—proving, perhaps, that retro works best on small cars when it can add that all-important sentimental touch.

PININFARINA

Pininfarina is to car design what Mozart is to music; a multi-talented master of elegance and refinement; a determined innovator when boundaries need to be pushed; and a resolute populist in bringing widespread access to high-class design through volume brands at accessible prices. Pininfarina is car design, and the history of the company is in many senses the history of the whole car industry.

Pininfarina's most widely lauded designs have been for Ferrari, serving a narrow elite of the rich and famous and setting the agenda for other luxury manufacturers. But where its influence was most profoundly felt was in the mass market, where its fresh, clean, and elegant designs for volume producers such as Alfa Romeo, Peugeot, and Austin in the 1950s raised the bar for style. Throughout the 1960s and through to the highly successful 306 and 406 in the late 1990s, Pininfarina's designs brought distinction to the Peugeot lineup, with Alfa Romeo maintaining the momentum into the new millennium.

Right from the earliest days, Pininfarina's steady stream of groundbreaking concept cars and glamorous Ferraris has served to bring new ideas into circulation. The two-seater Cisitalia 202 of 1947 stunned the automotive world with its low build and smooth, uncluttered contours; the 1965 mid-engined Ferrari Dino Berlinetta Speciale concept introduced a radically new aesthetic for exotic sports cars, including the production Dino 246 and all the later 308/328 family, while the ultra-aerodynamic CNR "banana car" study of 1979 set the entire industry on a course to reduce wind resistance and save fuel.

What made Pininfarina successful when so many other aspiring studios failed was the combination of creative flair and a sharp business acumen that allowed it to move with the times and on occasions anticipate industry trends to gain commercial advantage. Its leading light, Battista Farina, nicknamed "Pinin," transformed the business from a maker of one-off, coach-built special designs for rich people in the 1920s to a much bigger builder of small series models in the 1930s. In the chaos that followed World War 2, Battista realized that the market for extravagant luxury cars had disappeared, so he turned the firm over to the production of smaller cars based on Fiat and Alfa Romeo chassis, with the charming Giulietta Spider as its proudest achievement.

Many decades later Pininfarina would become weighed down by its manufacturing overheads as outside carmakers such as Peugeot, Lancia, Mitsubishi, Ford, and Fiat pulled production contracts back in house; this prompted a major realignment as a design and engineering consultancy operating across multiple sectors and disciplines, though the flow of trendsetting concept cars and new designs for Ferrari continued unabated.

Yet even this was not enough to stem the post-crisis financial tide as automakers scaled back on contracted-out design work. In common with other studios such as Bertone and Zagato, Pininfarina was close to collapse: it was acquired by India's Mahindra Group in late 2015 but continues to operate from its headquarters near Turin in Italy.

GLOSSARY: THEMES, STYLES AND TRENDS

Aerodynamic: Style of design where characteristics suggesting the smooth flow of air are emphasized. The discipline of aerodynamics takes in stability and downforce as well as the minimization of wind resistance.

Art Deco: Artistic and architectural movement that influenced the auto industry, especially craft-level coachbuilding designers, in the 1920s and 1930s.

Bauhaus: Germany-based movement that brought the value of "form follows function" to industrial design and architecture in the 1920s.

Boxes: The visual elements, volumes, or masses that make up the basic shape of a vehicle. Thus sedans with projecting trunks are three-box designs, hatchbacks are two-box, and most minivans are one-box, often referred to as monovolume.

Cd: Coefficient of drag, or how easily the vehicle's shape allows it to slip through the air. The lower the figure the better: anything below 0.22 is very good.

CdA: The Cd figure multiplied by the vehicle's frontal area, giving an indication of the wind resistance the vehicle faces.

CFD: Computational Fluid Dynamics, or the use of computer techniques to map airflow over—and under—the vehicle to develop the optimum aerodynamic shape.

Crossover: A vehicle based on a car platform but with the taller stance and elevated driving position of an SUV. A major growth segment after 2010.

Folded Paper: Informal name for a style of design pioneered by Giugiaro in the late 1960s, combining clean lines, flat surfaces, and sharp angles between panels.

Footprint: The area of road space occupied by a vehicle, i.e. length multiplied by width.

Fuselage: For aircraft, the main body. In older cars with separate fenders, the main body enclosing the engine and passenger compartment; now sometimes employed to describe the whole body, especially if it is low and curved in cross-section.

Hot Hatch: A term that originated in the 1970s to describe sporting versions of standard family hatchbacks, often distinguished by wider wheels, lowered suspension, larger or multiple exhaust outlets, and the use of blacked-out panels and trim.

Kamm Tail: Aerodynamic feature where the tail is cut off vertically to control the air flow, giving the effect of a longer rear section.

Ladder Frame: Traditional vehicle construction principle where a rigid chassis carries the engine, transmission, suspension, and wheels, with the non-stress-bearing body mounted on top. Now confined to commercial vehicles and some pickups and SUVs.

Microbus: Volkswagen trademark for a compact, bus-like vehicle based on a car platform and seating multiple passengers in two or more rows.

Microvan: Small urban van, often with forward control and the engine mounted under the floor. Originated in Japan but now popular in many markets.

Minivan: US term for multi-seater vehicle aimed at families and leisure activities, generally with multiple rows of seats that can be folded or removed.

MPV: Multi-purpose vehicle, see *Minivan*.

Modernism: A movement in 20th century architecture that sought to simplify forms and apply industrial principles to provide quick, effective, and attractive structures. In automotive applications, it is perhaps best expressed in quality volume production brands such as Volkswagen and Toyota.

Packaging: The art and skill of fitting people, luggage, and mechanical elements within a given vehicle envelope. Good examples are the Toyota iQ and the 1959 Mini.

Pontoon: A mid-century body style where the front fenders were extended rearward to merge with the rear fenders, creating a smooth side and eliminating the running boards in the process.

Product Design: A design ethic that prioritizes function over form but maintains an elegant simplicity, avoiding automotive clichés and preferring rounded, soft graphics.

Retro: Philosophy that refers to historical models for cues or inspiration, often aiming to evoke the emotions associated with fondly remembered old designs.

Retrofuturism: A term intended to place a positive, technically focused slant on retro design. Employed by designer J Mays, a leading advocate of retro design while at Volkswagen and, later, Ford.

Streamlining: Design philosophy aiming to minimize wind resistance and present a smooth, flowing appearance. Contrast with *aerodynamics*, above.

SUV: Sport Utility Vehicle, a catch-all description encompassing dedicated off-road vehicles as well as high-riding models intended mainly for on-road use.

Unitary Construction: Method of vehicle construction where the body and floor are welded together to form a strong and safe box structure. Much lighter and more rigid than body-on-frame construction.

Wedge: A dynamic style of design that became popular in the 1970s as awareness of aerodynamics grew. Inspired by Marcello Gandini's Bertone Carabo concept car of 1968 and popularized by British Leyland models such as the Princess and Triumph TR7.

4756

960

R 15

3640

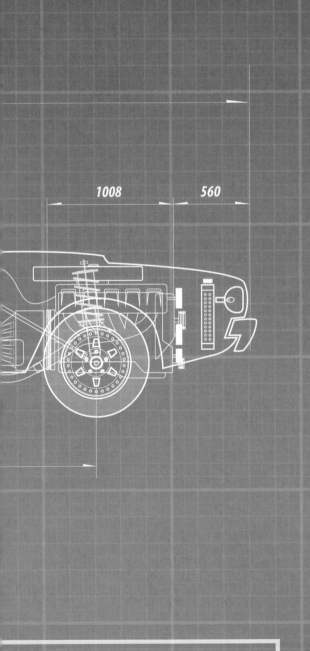

1008   560

INNOVATION

INTRODUCTION
PUSHING THE BOUNDARIES

Some industries progress in smooth, barely discernible, incremental steps; others, such as computers or electronics, advance in a series of dramatic leaps and bounds, where everything that went before suddenly becomes obsolete.

With cars, it's a blend of both. Those true quantum jumps in style and substance, the ones that instantly feel like a leap into the future, could perhaps be counted on the fingers of one hand. They are the genuine automotive landmarks that drew gasps of astonishment when first revealed, waypoints that triggered a seismic shift in automotive perspectives. Among these, the Citroen DS is the clearest candidate of all. In the drab, rectilinear atmosphere of 1955 it came as an astonishing revelation, a vision of a future world that pushed every boundary with its radical streamlining, its advanced engineering, and the overall audacity and coherence of its design. Clearly, nothing would ever be the same again.

The BMC Mini, launched just four years later, would turn automotive engineering upside down with its revolutionary super-compact transverse engine, but its prosaic style and everyday mission initially prevented it from capturing the public imagination in the way of the DS or, in 1961, the blatantly seductive Jaguar E-Type. The Jaguar, inspired by the racing D-Type, left every onlooker transfixed in open-mouthed admiration: it pressed buttons that had never been pressed before. So too did the NSU Ro80 of 1967, stunning audiences as the ambassador for a futuristic aerodynamic era as much as for its rotary engine. Just the year before, new-boy Ferruccio Lamborghini and designer Marcello Gandini had rocked the sports car establishment—and rattled avowed rival Ferrari—with the spectacularly low-slung shape of the mid-engined Miura, the car that gave birth to the whole supercar class.

Those may be some of the clearest landmarks in the automobile's 130-year story, but countless others deserve mention too. Fresh shapes with such a big impact that they kicked off new styles or prompted a wholly new typology of vehicles; or new designs where the innovation is under the skin, like the Toyota Prius as the first mass-market hybrid, or the Nissan Leaf as the first electric family car.

High up on the style roll call are the Ford Mustang, which in 1964 invented the muscle car genre; the Renault 16 (1965) and 5 (1972) as the first family and urban hatchbacks, respectively; the 1970 Range Rover as the archetype of the luxury SUV; and the 1984 Renault Espace as the first minivan with genuine style as well as the instigator of the one-box silhouette. Renault scored again with the mid-size Scenic minivan in 1995. In with a shout, too, are Toyota's 1997 RAV4, which introduced the notion of a small fun-driving 4x4, and the Nissan Qashqai (Rogue Sport in the US), the first volume-selling family crossover. Also worthy of mention are the Mazda MX-5 (1985), which singlehandedly revived the compact roadster class, and the Tesla Model S, which in 2012 encapsulated the future of driving as the first truly credible zero-emission luxury car; it may turn out to be the most influential turning point of all.

INNOVATION
CONCEPT CARS I: SETTING THE IMAGINATION FREE

BUICK Y-JOB

This 1938 two-seater is regarded as the auto industry's first true concept car. It pioneered the horizontal grille and ideas such as concealed headlamps, power windows, and a convertible top.

DODGE TOMAHAWK

One of the most outrageous concept vehicles ever developed, the 2003 Tomahawk consisted of a 500-horsepower V-10 engine running the length of a motorcycle-like frame, with twin wheels front and rear. It succeeded in attracting publicity and several were built as special orders.

RENAULT DEZIR

Renault design was at a low point when Laurens van den Acker joined as design chief in 2009, he set out to provide a fresh brand philosophy, and the 2010 DeZir concept for an electric supercar was given an enthusiastic welcome. Its distinctive frontal identity is now used on all Renault production cars.

They have been described as the shooting stars of the car world, burning brightly and illuminating everything around for a few brief instants before fizzling out and disappearing from sight; others see them like a shop window or a trailer for an upcoming new movie. But whichever way they are viewed, concept cars have a vital role to play in bringing new and radical designs to public attention—and perhaps even into production.

Automakers develop concept cars for a range of reasons: to gauge public reaction to new types of vehicles; to float new design themes prior to the launch of an upcoming model; to boost corporate prestige and morale; or to make a statement of intent about future policy—such as Bentley presenting its EXP 12 concept for an electric sports car.

But there is another, wholly different strand to concept car design: the automotive fantasy, the dream car incorporating wildly imaginative ideas that may not yet be technically viable. Here is where designers are let off the leash and set free to give shape to their most inventive and most fanciful thoughts; these cars will never take to the road, but they will plant a seed that may take root and blossom in years to come.

Designers love working on concept cars because there are no rules: no safety standards, no need to be practical or sensible—or even to provide room for an engine. The job is to inspire, to create an aura of excitement and futuristic expectation around the brand.

Past masters at the art of putting fantasy onto wheels include GM's legendary Harley Earl and Bill Mitchell, with their series of astonishing, aircraft-inspired concepts that toured with the firm's Motorama roadshow in the 1950s. Pininfarina and Bertone, too, have delivered many game-changers over several decades, while in 1978 Giugiaro presented the visionary high-riding Megagamma concept that went against every trend of the times but led directly to stylish minivans such as the Renault Espace and the compact Fiat Uno.

Concept cars can present engineering ideas too, as shown to dramatic effect by GM's Autonomy concept of 2000, its skateboard-like chassis demonstrating the compact packaging of a hydrogen fuel cell powertrain.

CONCEPT CARS II: FROM SHOW STAND TO SHOWROOM

CORVETTE STARTED AS A CONCEPT

Seven generations of America's best-loved sports car, the Chevrolet Corvette, began with a hasty fiberglass-bodied show car wheeled out as part of General Motors' 1953 Motorama travelling roadshow.

HYUNDAI CONCEPT THAT ANTICIPATED BMW X6

SUVs were all the rage in 2006, leaving observers puzzled by Hyundai's HCD9 Talus concept. With a sleek coupé roofline atop a muscular body on oversized wheels, it seemed an incongruous set of mixed messages. Two years later, BMW's X6 coupé off-roader appeared; now, such proportions are common.

THE MONSTER CADILLAC THAT DIDN'T MAKE IT

After Cadillac celebrated its centenary in 2002, the 2003 Sixteen concept was intended as a spectacular finale. Over 5.7 meters in length and with a V-16 engine, it proved an embarrassing extravagance for a company already in big trouble.

We described earlier how even the most way out of concept car ideas, with no possibility of ever going into production, can still play an important role in an automaker's strategy. But concept cars can also be used to manipulate public opinion in a different way: to soften up potential buyers' attitudes toward upcoming models, especially if there is a big change on the way.

The Nissan Qazana prototype was a polarizing design and shocked many when unveiled at the 2009 Geneva Motor Show. But Nissan knew that the production Juke was waiting in the wings; it, too, was a polarizing design, but the Qazana served to prepare the public for the shock that was to come. In a similar vein three decades earlier, Ford's ultra-aerodynamic Probe III concept of 1981 paved the way for the then-radical Sierra the following year.

Some concepts may not be fully feasible for production but do carry a hint of plausibility, perhaps by being built on an existing engineering platform. Two examples from Audi prove the point: the TT sports coupé concept of 1995 was greeted with such passion that by 1998 it had become a successful production car; the 2000 Steppenwolf concept, on the other hand, was met with nothing but bafflement. Its mission of being a "mountain bike on four wheels" was not understood at all—until some years later when coupé-like off-roaders such as the BMW X6 and Range Rover Evoque became big hits in the market and Audi stepped in with its own Q-badged models.

The story of the Renault Scenic is also illustrative. The distinctively shaped concept in 1991 floated the idea of a compact but warm and inviting multipurpose vehicle with a highly imaginative interior and storage solutions; when the production model appeared in 1996 it had a different look but retained the practical brilliance and a streak of innovation. It was a massive hit and created a whole new market segment. More recently, the fourth-generation Scenic was previewed in 2014 by a much-applauded concept, the R-Space, and in 2015 Porsche's low-slung and harmonious Mission E set the scene for the sports car maker's forthcoming electric sedans.

INNOVATION
EXCESS ALL AREAS

AND THE LONGEST CAR IS. . .
Today's chauffeur-driven Mercedes Maybach Pullman measures 6.5 meters (21.3 feet) and just beats the 1975 Cadillac Fleetwood 75, at 6.4 meters and also a chauffeur vehicle.

HUMMER HYSTERIA
In the early 2000s the military-derived Hummer H1 and its civilian counterpart H2 became popular with celebrities thanks to their size and bullying demeanor. Despite their intimidating width and height, both models are shorter than full-size US sedans and SUVs, which top out at 5.6 meters.

AERODYNAMIC DEVICES
Some of the best examples come from the US muscle car boom in the 1970s and the European hot-hatch wars of the 1980s. Leading contenders include the Plymouth Super-bird, with its crazy, towering deck-lid wing, and Ford's Sierra and Escort Cosworth with dramatic double-decker rear aerofoils.

Many designers take pride in pushing the limits in style, creating shapes that unsettle or polarize—whether through proportion, surface language, or unconventional detailing. Others prefer extremes of scale, of speed, of ostentation—or combinations of all three. It is not for nothing that the 1959 Cadillac Eldorado Biarritz, measuring nearly 19 feet (225 inches, 5.72 meters) from its ornate chrome-laden nose to the apex of its massive pointed tailfins, is one of most memorable car designs of all time.

Though the late 1950s marked high water for US automotive excess, full-size cars continued to hover around the 5.7-meter point for a decade or more, and in the 1980s the growing ranks of SUVs swelled to a similar size—but of course taller, heavier, and more imposing. Today, full-size SUVs such as the Ford Expedition and Chevrolet Suburban are nothing out of the ordinary in North America, and their design language helps emphasize their stature.

In fact, proportion is more important than sheer size, as Mercedes discovered when it launched Maybach as a top premium rival to Rolls-Royce, which had recently been revived by BMW. The long and low 5.7- and 6.2-meter Maybachs looked out of proportion, as if a crude stretch of an existing model, the S-Class. BMW, for its part, insisted on a high and mighty stature for its reborn Rolls-Royce Phantom and, at 5.8 and 6.1 meters, respectively, both standard and extended wheelbase versions looked well proportioned. All these, however, are made to look tiny alongside the prewar Bugatti Royale, 6.4 meters from stem to stern.

Sports cars have an easier time making a powerful impression. Low build, extreme width, dramatic proportions, and outrageous aerodynamic devices all spell out horsepower, technology, and speed. Lamborghini is the real master here: its supercars have shocked, stunned, and amazed ever since the Miura and, especially, the 1971 Countach. Today, that aggression is more nuanced, but the marque's sharp surface language is more distinctive than that of rivals Ferrari, McLaren, and Porsche. Memorable? Certainly—but not in quite the same way as a '59 Cadillac.

INNOVATION
GREAT IDEA, BAD TIMING

MAXI OPPORTUNITY
Austin's 1969 Maxi was a roomy hatchback and could have been as big a success as the VW Golf, which came five years later. But it proved unreliable and troublesome from the start and developed a poor reputation.

CORVAIR'S CHALLENGE
Compact cars were rare in the US in the 1950s but Chevrolet broke with convention and in 1960 launched an innovative model with a rear-mounted, air-cooled, flat-six engine. The Corvair would have been a hit had it not been for its tricky handling, which gave Ralph Nader the title for his 1965 book *Unsafe at Any Speed*.

WAS TOYOTA'S IQ TOO SMART?
Tiny proportions and small engines, plus decent ride, handling, and safety. The clever Toyota iQ was little bigger than a two-person Smart but sat four—just. It deserved to succeed yet somehow failed.

We all celebrate innovation, but sometimes ingenious new ideas are not the right approach. However bold the invention might be, it will not succeed until customers are ready for it—or actually need it.

In 1994 Volkswagen offered an early and crude form of the engine-stop systems that are now commonplace: they failed because no one at the time wanted stop-start, and now no one remembers it. In 2000, Audi presented its A2, the world's first aluminum spaceframe small car: very light and very frugal, it was a flop because it was too clinical and cold, and buyers headed toward Mini showrooms for fun and glamor. But had the A2 been launched five years later, it could have been a hit.

Half a century earlier, Preston Tucker in the US was putting the finishing touches to his advanced rear-engined Torpedo sedan, one of the first cars to consider crash safety. The project collapsed amid allegations of fraud and conspiracy. In the UK Jowett presented its Javelin in 1947, a sophisticated, streamlined six-seater with a flat-four engine and a talent for winning rallies. But it too failed, both mechanically and commercially.

Others fall at the first hurdle because they don't give a clear message to potential customers. Was the Mercedes R-Class (2005–2012) a station wagon, a minivan, or an SUV? No one really knew, and sales never picked up. Yet Volvo had a huge hit with its XC90 in exactly the same sector. More celebrated is Renault's Avantime, launched in 2001 amid heated debate over its strange proportions, tall minivan-like build, cleverly hinged doors, and four-seater cabin. Fewer than 8,000 were sold—mainly to trendy urbanites—before its builder collapsed in 2003. A parallel project, the Vel Satis, was a brave tilt at the German-dominated executive car market but proved too quirky for its target clientele; it lasted only slightly longer.

The moral is clear. Creativity is the lifeblood of the car business, but innovation must be carefully prepared to have a chance of success. Former Renault design director Patrick le Quément once said that the biggest risk was not to take a risk—a sentiment every car designer and car enthusiast would happily endorse.

INNOVATION
UGLY DUCKLINGS, LEMONS, AND LOST CAUSES

PONTIAC AZTEK

The 2001 Pontiac Aztek is the ugliest car of recent times. It offends against almost every rule of design, from its clash of lines and slabby planes, its split-personality front, child-like detailing, and its block-like proportions.

MINI SHOULD KNOW BETTER

Everyone loved BMW's reborn 2001 Mini, but as the generations rolled by it became bigger, fatter, and a caricature of the original. And with each derivative, the proportions suffered as the designers struggled to maintain their theme.

MERCEDES-BENZ GLK

Without a three-pointed star on the grille of this 2008 debutante, few would have detected the hand of Mercedes in its design. The blocky, ungainly SUV aimed to convey the toughness of the military-style G-Wagen in the compact luxury sector but lost out to Audi's finely crafted Q5.

Cars born without the blessing of good looks are the outcasts of the automotive world. They are vilified in the press, mocked on social media, and ridiculed whenever people gather to talk cars. They stall in the showrooms too—but not all of them deserve to fail.

Some, despite their misshapen appearance, have hearts of gold and are great cars underneath—the duck-ugly Fiat Multipla is the tediously cited example—while others like the Chris Bangle-designed Fiat Coupé and BMW's Z3-based M Coupé had an initial shock value that gradually softened into positive appreciation. A third category where things can go wrong is when designers try to produce a bigger car out of a small one. Sedan and wagon spinoffs of hatchbacks are a case in point: the lofty Toyota Yaris Verso is ugly but irritatingly practical, as is Skoda's Roomster.

Historical horrors abound, with British Leyland, inheritor of Austin, Morris, and a multitude of English marques, accounting for more than its fair share. The cheddar-cheese wedge of the Triumph TR7 coupé, the similarly wedgy Princess, and the blob-like Allegro all braved a new style, and all missed the mark. Ford's missteps include the "jelly mold" Sierra and the final iteration of the big Scorpio, where it never recovered from the grafting on of a gaping, fish-like chromium mouth and an enormous unbalancing trunk. North America's Ford Taurus was similarly afflicted.

Yet, more recently, the openly confrontational Nissan Juke has polarized opinions but is selling like hot cakes to fashionable city dwellers. One enduring mystery is why super-ecological models from Japanese manufacturers have to be so over-complicated and unsightly. The first Toyota Prius was modest and plain; today's fourth-generation model is full of sharp angles, creases, and strangely unbalanced surfaces, and its hydrogen-powered relative, the Mirai, is positively un-coordinated from almost every angle. As for Honda's Clarity, rival to the Mirai, its lumpy, oversized body appears about to swamp its modest chassis. If this is the way eco car design is heading, we are in for an ugly future.

CITROEN

Whenever it comes to innovation and inspiration in automotive design, Citroen is the name that first springs to mind. From its origins in 1919 to the takeover by Peugeot in the 1970s, Citroen hatched brilliant new ideas in every area.

Front-wheel drive, all-in-one construction, hydropneumatic self-levelling suspension, streamlining, the single-spoke safety steering wheel—all and much more were pioneered or popularized by Citroen. So, too, were a highly distinctive treatment of external style and wildly imaginative approaches to the interface with the driver—controls, instrumentation, and the way the cars feel on the road. Above all, Citroen cars were truly different—the standout shapes that appealed to initiates and creative types, people who bought in to an open-minded, left-field philosophy that dared prefer originality to conformity.

Citroen was the author of two of the greatest step changes in automotive design—the Traction Avant in 1934 and the DS in 1955. Both were brave and brilliant advances born out of an urgent creative impetus that refused to accept the status quo. Yet each came at enormous cost: developing the Traction bankrupted André Citroen's company and the attendant stress cost him his life the following year; and though the DS gave substance to Citroen's golden age in the 1950s and 1960s, its success sowed the seeds of an almost drug-like addiction to engineering adventure that would again bring the company to its knees in the 1970s.

The humiliating rescue by an archrival, the conservative and prudent Peugeot, kept the Citroen name in the showrooms by dint of the hasty application of the Citroen logo to sensible Peugeot models. This horrified the legions of Citroen loyalists and touched on the dichotomy that has come to characterize the marque ever since: the struggle between the eccentric and often costly creativity implicit in the brand's genes, and the accountants' call for simplicity, rationalization, and the kind of conformity that sells easily and turns a profit.

So the past four decades of Citroen's history have witnessed several low points in the shape of warmed-over Peugeots rebranded with the double chevron, balanced by occasional flashes of brilliance: the Gandini-styled BX hatchback that dared do battle with the VW Golf, the eccentric C6 luxury sedan, and the funky C4 Cactus.

But for those who worry that the groundbreaking spirit of Citroen may have been diluted out of all recognition, there could be an answer in store: in 2010 Peugeot Citroen lent the name DS to its freshly-launched aspiring premium brand. Implicit in that decision was an honoring of the universally celebrated 1955 archetype, as well as a promise to live up to the original's spirit of innovation. And on the basis of today's colorful DS3 and serenely sophisticated DS5, that promise could hold good.

Aerofoil: Strictly speaking, the cross-section of an aircraft wing that supplies lift when air flows over it. Applied upside down on a car to provide downforce and stability.

Air Cooled: An engine where the cooling medium is air, not water. Now found only on motorcycles.

Battery Electric: An electric car powered solely by batteries, such as the Nissan Leaf or Tesla Model S.

Compact: US auto classification, introduced in the late 1950s, for cars slightly smaller than the standard or full size.

Concept Car: One-off prototype or show car developed by designers to showcase new styling themes or configurations. Not necessarily powered or mobile.

Design Language: The way a designer handles surfaces and their intersections. Can be smooth and soft, or taut and geometric, simple or complex. Also referred to as form language or surface language.

Double Chevron: Trademarked emblem of Citroen, taken from the meshing of herringbone gears.

Flat Four, Flat Six: An engine where the cylinders are arranged in separate rows of two (or three), lying opposite one another on either side of a central crankshaft.

Fuel Cell: A propulsion system that mixes hydrogen fuel and oxygen from the atmosphere to produce an electric current and power a traction motor. The fuel cell has no moving parts and its only emission is water.

Full Size: US vehicle size classification which used to be based on length (around 220 inches, or 5.6 meters) but which is now determined by cabin and trunk volume.

Hybrid: Vehicle power unit that combines electric and combustion-engine elements.

Hydraulic Brakes: Brakes that are operated by the pressure of fluid in a system of tubes and cylinders; began replacing cable-operated brakes in the 1930s.

Hydropneumatic: Suspension system developed for the 1955 Citroen DS using high-pressure fluid and air to provide a smooth ride and self-levelling to control pitch.

Muscle Car: US market category inspired by the Ford Mustang in 1964 for high-powered (generally V8) compact sports cars with assertive looks and extrovert style.

Platform: A predefined set of vehicle components, including chassis structure, suspension, brakes, and electrics, that can be standardized and built in large numbers to minimize unit costs.

Pullman: Originally, a railway carriage with an exceptional level of comfort. Used by Mercedes-Benz for the long-wheelbase version of the 1963 600 *Grosser Mercedes* and now revived for the Maybach version of the S-Class.

Rotary: An internal combustion engine, developed by Dr. Felix Wankel, which uses a triangular rotating piston rather than individual pistons going up and down.

Spaceframe: Type of car construction where the structure is made up of a framework of small loadbearing members, covered by an external skin. Can be lighter than some conventional structures.

Stop-start: A system that automatically switches the engine off when the vehicle is stationary, such as at a red light, and restarts it when the driver selects gear or presses the accelerator.

Supercar: Magazine term coined in the 1960s to describe the new wave of highly powered exotic sports cars, generally mid-engined, initiated by the Lamborghini Miura.

Transverse: The placing of the engine with its crankshaft across the vehicle's width, to improve space utilization. Pioneered by the BMC Mini in 1959 for front engines and by the Lamborghini Miura for centrally mounted engines.

Wheelbase: The distance between the front and rear axles, and a key determinant of the amount of passenger space in the cabin.

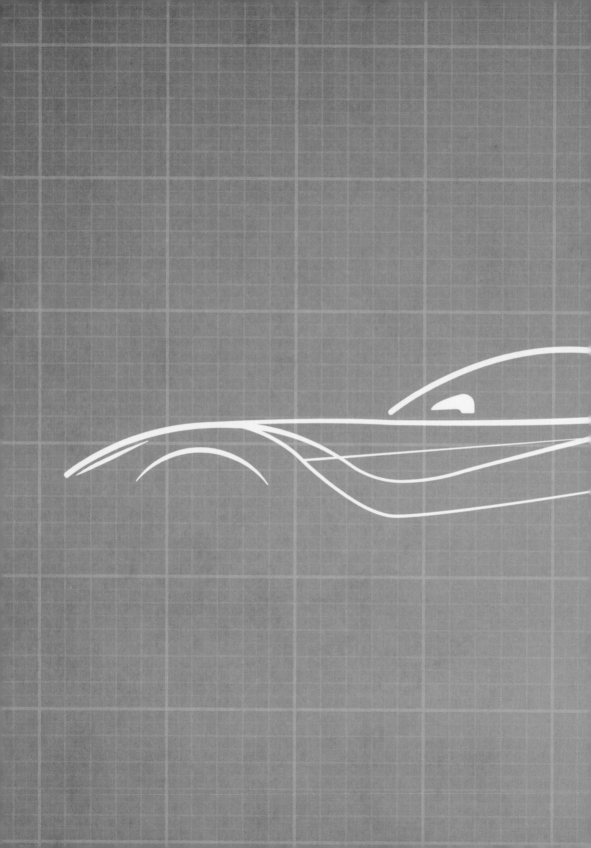

THE ELEMENTS OF STYLE

INTRODUCTION
STYLE, SUBSTANCE AND STATURE

It's true for people, and it's just as true for cars: some have real style, while others clearly don't. Style is noto-riously tricky to define, yet everyone knows it as soon as they see it.

Cars are objects that play upon our subconscious to stir up strong emotions and make that sense of style particularly acute. People care about cars in a way that they don't when it comes to other mass-produced goods like coffee makers or computers. A car that gets it right can exert a mysterious magic that locks the observer in an admiring embrace; one that gets it wrong will leave that same person cold and unmoved. Yet, paradoxically, both are products shaped from the same raw materials, sharing similar engineering and, ultimately, serving the same purpose.

That day-and-night difference is down to that one elusive, indefinable quality: style. It's the magic dust that a good car designer understands intuitively and it can emerge from the pencil strokes of the first sketch. It's to do with first impressions, but also lasting impressions—a truly classy design will continue to reward the onlooker with visual interest for a long time.

First impressions generally begin with the car's stance and stature—its relationship to the road, whether it is low and stably planted on the ground or, at the opposite extreme, taller and more likely to topple. The overall proportions count too: how the front and rear volumes balance the visual mass of the passenger compartment in the center. Implicit in this is the "gesture," the dynamic graphic the car's profile makes: is it homely and accommodating, upright and stately, or wedge-shaped to convey forward thrust? Could it even be openly aggressive in the manner of a predatory animal, crouched and ready to pounce?

Once those first impressions have been registered, what was basically a silhouette will reveal more and more as the observer comes closer: the overhangs front and rear, the size and fit of the wheels within the body sides, the proportion and the profile of the glasshouse in relation to the lower body, and the way the designer handles the shaping of the body's surfaces to control the interplay of light and shade. Each of these helps bring the design to life and build the initial picture that will determine the impact of that first glance.

In the sections that follow, we go behind the studio doors to discover the tricks of the designer's trade, the special skills that shape customers' impressions and manipulate their emotions to generate that all-important flush of "I want one" passion that we all know and love.

THE ELEMENTS OF STYLE
HOW TO READ A DESIGN

LEXUS NX EXAGGERATES PROPORTIONS
Even when it comes to SUVs, designers prefer low and wide. But such is the entertainment value of Lexus's complex and jagged surfacing on its NX crossover that its tall and narrow, slab-sided proportions tend to get overlooked.

THE BIG STRETCH
For an illustration of how a design's side window profile can stretch a car's impression, check out the Mercedes CLS shooting brake. The chrome surround stretches in an enormous arc from screen pillar to taillights, visually lengthening the whole vehicle.

PERILS OF THE RUSSIAN DOLL APPROACH
Premium marques often strengthen brand identity by applying a similar style to models in each size bracket. But Audi's SUV template, which works well on the premium Q7 and mid-sized Q5, comes unstuck with the smaller and too tall-looking Q3.

The character of a vehicle at its most fundamental level is determined by its overall proportions—the relationship between its length, height, width—and also the size and positioning of its wheels within that length. But things may not always be exactly what they seem, and designers have plenty of tricks up their sleeves to fine-tune the message they wish to convey.

A shorter design can be made to look longer by skillful emphasis of horizontal lines, perhaps with a character line running the full length of the vehicle; a low build helps too, as does a "floating" roof that highlights the horizontal elements. What works less well is to add overhang front and rear. This makes the design look overbodied and clumsy, though, again, tricks such as the rounding off of the front and rear corners can help.

These relationships, together with the profile of the roof structure or glasshouse, determine the inherent attitude of the car. Wedge profiles with a rising waistline, as we have seen, convey a dynamic forward thrust, even on a tall SUV—as demonstrated by the Range Rover Evoque. A level waistline gives a calmer impression, while a "fastback" roofline falling away to the rear signals a coupé-like sportiness and is currently much in vogue to give tall crossovers an energetic look that belies their bulk.

Nevertheless, many would argue that the single most important element in a vehicle's visual signature is the glasshouse, or window area; allied to this is the proportion of side glass in relation to the painted metal of the body. In recent years the trend has been toward a tougher, more secure look, with shallower side windows. Suitably highlighted, a well-shaped DLO (the shape made by the side glass) can visually stretch the length of the whole car—just look at the Jaguar XJ.

Stance, finally, is one of the most important subliminal messages telegraphed by a design. Low and wide, especially with the wheels set as wide as possible, signals sporty stability; tall, and with plenty of suspension travel for chunky tires in oversized wheel arches, shows terrain-crossing prowess; and a long wheelbase in relation to waist height gives a sense of space and luxury.

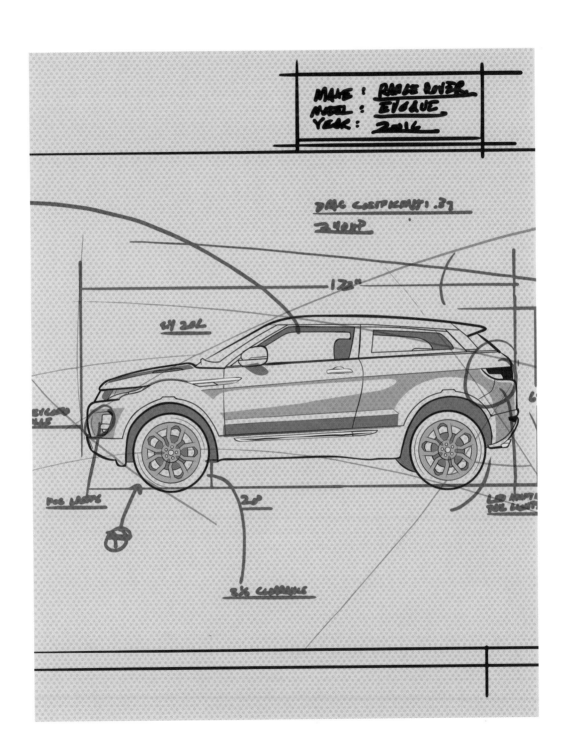

GIORGIETTO GIUGIARO—THE MASTER

If the history of Pininfarina is a close parallel with the history of car design itself, then it is Giorgietto Giugiaro who provides the master class. Over the course of nearly seven decades he and the Italdesign studio he founded have designed more than 200 production cars and over 100 show concepts; his designs have sold in huge numbers—60 million at the last count—and in 1999 he was voted Designer of the Century.

His sure touch with both styling and the underlying engineering and packaging made his designs attractive, influential, and, most importantly, commercially successful. He is most famous for having penned the original Volkswagen Golf, launched in 1974; its successors helped propel VW to the world number-one slot in 2016. Other huge hits signed by Giugiaro include the Fiat Panda, Uno, and Punto, as well as the Alfasud and 1979 Lancia Delta, which on several levels is a more satisfying design than the VW Golf.

In each of these models there was a crispness, freshness, and clarity of design absent from rivals. Giugiaro did more than any other designer to democratize good design through high-volume products, helping raise everyone's game in the process. But he has also been immensely influential with his more visionary concepts, most notably the 1977 New York Taxi and the related Lancia Megagamma the following year; together, these proposed the tall, space-efficient configuration that went on to become the family minivan and also inspired the Fiat Uno in 1983. Earlier, the spectacular Maserati Boomerang concept had crystallized Giugiaro's sleek but controversially sharp "folded paper" design language, an approach that yielded, among others, the 1974 Lotus Esprit and the celebrated BMW M1 of 1978.

The M1 was not Giugiaro's first work for BMW, however. Born into a family of artists in 1938, he went to art school in Turin, where his car sketches were spotted by the legendary Fiat designer Dante Giacosa. Taken on as an apprentice in advanced design at Fiat, Giugiaro was later spotted by Nuccio Bertone. Industry lore says that Bertone sold Giugiaro's test piece to Alfa Romeo, where it became the 2600 coupé; certainly, the elegant BMW 3200CS coupé he completed at Bertone in 1961 shares its many of its most attractive genes with the big Alfa.

After a short spell at Ghia in 1967, during which he designed the influential De Tomaso Mangusta, Giugiaro teamed up with Aldo Mantovani to found Italdesign, the studio that would set the industry's design agenda for the next 40 years. In this time Giugiaro was able to branch out into a much wider variety of products such as cameras, watches, sewing machines, furniture, and even pasta shapes. But after 2010, when Volkswagen bought a 90-percent stake in Italdesign, Giugiaro's influence began to wane.

Five years later, he quit his own company by selling his shares, but he still stands as the most talented and most prolific designer the car world has ever known.

THE ELEMENTS OF STYLE
MASTERCLASS: GIUGIARO'S GREATEST HITS

WRONG CALL ON GOLF
It's hard to believe, but many senior managers at Volkswagen in the early 1970s believed the Golf, scheduled to launch in 1974, could not work. Fortunately for VW, they lost the argument.

GIUGIARO'S MONSTER MACHINE
Not every Giugiaro design was elegant and attractive. The giant Columbus concept of 1992 was an extravagantly luxurious six-meter road yacht seating seven on two tiers. It has yet to catch on.

MOVIE MAGIC
Alongside Dustin Hoffman's *The Graduate* Alfa, one of the most famous four-wheeled movie stars is the Giugiaro-penned DeLorean DMC12, better known as Doc Brown's time-traveling machine in *Back to the Future*.

Giugiaro's output of production and concept cars has been so prolific that a roll call of all his designs would fill several pages of this book. In the early days, much of his most striking work found critical (rather than commercial) success as it was created for smaller and more specialized sports car manufacturers or confined to the motor show circuit as design studies.

Giugiaro's early work is characterized by smooth and simple surfaces and, above all, clarity and elegance of proportions; good examples are the Gordon-Keeble GT and the timeless 2600 Coupé and Giulia GT (later GTV) for Alfa Romeo. The BMW 3200CS also belongs to this period, and its best points influenced BMW's later in-house 3.0 CSi.

By the mid-1960s he had added the Iso Grifo, Rivolta, Simca 1200S Coupé, and Maserati Ghibli to his tally, but it was only after founding his own studio, Italdesign, in 1968, that he really hit the big time with the crisp, fresh style that became his trademark. Memorable designs including the Bizzarini Manta, Maserati Bora, and Alfa Romeo Alfasud mark Italdesign's earlier work, while his portfolio of projects for Volkswagen, including the Passat, Scirocco, and Golf brought him international acclaim, and the Alfetta GT and Alfasud Sprint coupé seduced sports car fans.

The radical Lotus Esprit introduced the "folded paper" style in 1975, echoed in milder form in the BMW M1 and DeLorean DMC12 and, in 1979, in the petite but perfect Lancia Delta.

The low-cost Fiat Panda, introduced in 1980, stayed in production for 23 years, and its success prompted the 1983 Fiat Uno, a groundbreaking small car with tall proportions. This, in many ways, was Italdesign's heyday, taking in the Saab 9000, Lancia Thema, Isuzu Gemini/Chevrolet Spectrum, the first SEAT Ibiza, the Lexus GS300, the Subaru SVX, and the Renault 19.

The following decades saw huge variety: the Fiat Punto, Bugatti EB110, Maserati 3200GT, Alfa Romeo Brera, Lamborghini Gallardo, and Fiat Croma, the first to explore a crossover-like architecture; as well as the Quaranta, Brivido, Clipper, Gea, and Frazer Nash Namir concepts.

THE ELEMENTS OF STYLE
WINDOWS, ARCHES, PILLARS AND DOORS

CAB FORWARD—
OR REARWARD

In the mid 1990s Chrysler made a big deal about its cab-forward design philosophy, which gave extra space for passengers. In 2004 BMW retaliated with just the opposite: its 1-Series had the cabin pushed right to the back to emphasize the model's long hood and rear-wheel drive, unique in the segment at that time.

CONCEPT RE-ANGLES IMAGE

What a difference a few degrees make: Audi's 2015 e-tron electric crossover quattro concept is tall and imposing, but elegant. The 2017 e-tron sportback concept, keeps similarly substantial underpinnings but its sloping fastback tail gives it a racy air.

FASTEST-EVER WINDSHIELD

In designer-speak a fast windscreen is one that is steeply raked backward. A leading contender for the fastest screen of all time is Bertone's 1968 Carabo show car, which kicked off the wedge-profile craze.

Proportions and ratios—especially the Golden Ratio—have been important influences in the worlds of art and architecture, and there is a surprising amount of geometry hidden in car design too. Angles, ratios, and forms subtly affect how we perceive designs, and seemingly trivial changes—say to the rake of a windshield pillar—can completely alter the character of a vehicle.

Two principal elements operate here: the wheels, and the pillars supporting the roof. In a harmonious, conventional design an imaginary line projected downward by the A-pillar (the one framing the windshield) often meets with the center of the front wheel; if that line projects forward of the front wheel, the car looks smaller and perhaps taller and more minivan-like. Further back gives a long-hooded impression for status and power—witness the Jaguar E-Type. The rearmost pillar is also a key signifier: the more strongly sloped it is, the sportier the vehicle seems, while vertical or near-vertical spells utility, like a van or truck.

Now project both of these imaginary lines upward to where they intersect above the roof. The lower the intersection point, the sportier the vehicle, and the front-rear location of that point gives a clue to where the visual center of mass lies. Above the B-pillar is the norm for family-type cars, rearward puts the stress on the spaciousness of the rear of the car, and forward emphasizes the front compartment. What's more, designers can manipulate these perceived angles with brightwork and masking to achieve almost any effect they desire.

The relationship of the wheelbase to the wheel size is another interesting metric. Concept cars generally have impractically large-diameter wheels, but sizes on production cars, too, have been growing in recent years as vehicles have become taller and their sides visually deeper. Larger wheels are necessary to offset the expanse of sheet metal, so even city cars now boast 15-inch rims or bigger. The close fit of the wheels within their arches is a key aim of good design and, as we will see in the next section, the design of the wheels themselves is far more important than might be imagined.

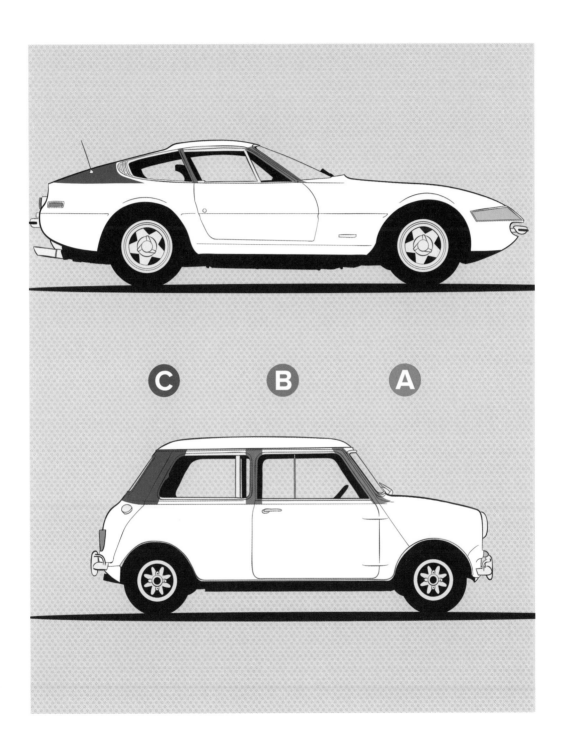

THE ELEMENTS OF STYLE
WHEELS

They are round, they are generally made of aluminum or some other cast alloy, and they are invariably tricky to clean. What is more, they are encircled by dull black hoops of rubber that are both damage prone and costly to replace. So why are wheels seen as such a crucial ingredient in a vehicle's design and style?

Firstly, and most importantly, the wheel is the only part of the vehicle that visibly moves when the vehicle itself is moving: it is the sole witness of motion, the sole signifier of speed. So that is why vehicles whose principal *raison d'etre* is speed—sports cars and performance models—tend to be specified with wheels that have strongly defined radial spokes or patterns; the clearly visible rotation of these patterns emphasizes the vehicle's motion and, therefore, potency. The chunky louvered alloy wheels of the original Saab 99 Turbo are a good example. They really move.

At the opposite end of the spectrum is the design of wheels for luxury cars. The essence of luxury travel is smoothness, serenity, and isolation from noise and outside elements that might betray a sense of speed. Luxury wheels therefore tend to be smoothly milled discs with little or no radial detail: these can hardly be seen rotating, allowing the vehicle to glide along gracefully with no visible sign of effort. The plain wheel covers of the ethereal 1955 Citroen DS illustrate the point.

The second key point about wheels and tires is that they symbolize the car's contact with the ground, and therefore its grip, braking and acceleration, and cornering. For this reason the wheel/tire combination must not appear too small for the vehicle's bodyside depth, and the tire must fit closely to the wheel arch that encloses it. Most importantly, the wheels must not be set too far in from the external bodywork or the vehicle will appear narrow and unstable. Manufacturers have gradually learned this lesson, and even entry-level models from volume brands now have wheels that, while relatively narrow, are still pushed outward to align with the wheel arch.

THE ELEMENTS OF STYLE
LINES OF BEAUTY

THE LINES THAT COUNT MOST

On some vehicle designs the most characteristic lines are those arising on the beltline—where the side glass meets the body's shoulder. Consider the controversial sawtooth line of the Citroen C2, or the clever and more alluring fall then rise of the 2016 Renault Scenic.

TWO-TONE DIVIDING LINE

The 1950s fashion for two-tone paintwork allowed designers to create entirely new types of lines on bodysides. Among the most graphic color splits were the Oldsmobile Rocket 88 and 98, along with the European Ford Taunus and Triumph Herald. Today Bugatti offers millionaires the option of multiple shades.

CHEVROLET SHOULDER LINE INSPIRES EUROPE

Ned Nickles's simple and elegant style for the 1960 Chevrolet Corvair proved unexpectedly influential. The horizontal shoulder line "rim" running the full perimeter of the car was copied by BMW, NSU, Simca, Hillman, Fiat, and others.

Lines could be regarded as the highlighters in the designer's pencil case. They come in many shapes and sizes: thick or thin, sharply defined or gently blurred. They are a vital tool in creating definition, marking transitions between planes, and in fine-tuning a composition to ensure the observer's eye is drawn in the right direction for maximum stylistic effect.

The most important lines are those that arise naturally out of a vehicle's three-dimensional form, perhaps signaling where the top of the fender rolls around into the bodyside; this transition could be smooth, giving a rounded profile and a soft line, or a tighter transition could result in a sharper, taut look. How this line is extended rearward, whether it stays level, rises toward the rear, or perhaps kicks up over the rear-wheel arches, will influence the stance and attitude of the whole design—hence the term character line, the single line that best defines the design. The "Coke bottle" cars of the 1960s and 1970s are good examples.

Creases or swage lines are generally inherent to the design too. They can be either raised or concave—a shallow linear indentation in the surface—and are used to add interest to flat panels. Many bring extra definition to a bodyside and are sometimes angled to draw the eye to, say, an air outlet. Bone lines, on the other hand, are always slightly positive and, as their name suggests, hint at the structure below.

Modern manufacturing techniques allow the pressing and painting of very tight radii on panels, enabling designers to specify much sharper creases than ever before. This is allowing the latest Audi and Volkswagen models a very taut surface treatment, with a powerful full-length character line undercut for added effect.

Too many competing creases and lines can lead to a confusing overall message, however. Hyundai, Kia, and Mercedes have all been guilty of this, and the latter's recent declaration of a "no more creases" policy shows how a clean and fresh impression can be achieved by simplified surface treatment and a strong, uninterrupted shoulder line.

THE ELEMENTS OF STYLE
LIGHT AND SHADE

LIGHT LESSON

Designs with simple, crisp lines best demonstrate the influence of light. For example, the first-generation Golf, with its clear, flat bodysides: below the side character line the surface is dark, but above it, the upper section always catches the light.

SURFACE FASCINATION

BMW's "flame surfacing" technique added three-dimensionality to almost every panel, catching a multitude of reflections in the process. It is shown to best effect in the Z4, especially in the detailing around the indicator light recesses on the front fenders.

THE INCREDIBLE DISAPPEARING PILLAR

Light, shade, and color effects can be exploited to alter the perceived proportions of a vehicle's structure. "Wrapping" a pillar in darkened glass can make it visually disappear, making the glasshouse look longer, and black insets can take the visual mass out of front and rear aprons.

It is the interplay between light and shade that helps bring a solid object to life. The highlights and the lowlights reveal the three-dimensionality of the object and the contours of its surface become clear. For designers, the advent of sophisticated software is allowing studios to build in more complex surface treatments to capitalize on the effects that light can bring.

Broadly speaking, the upward-facing surfaces of the vehicle—the hood, the roof, the tops of the fenders—will reflect the sky and thus appear bright; lower areas such as bodysides will, by contrast, generally come across as darker as they tend to mirror the ground—but not always.

Designers are clever in their manipulation of shapes and surfaces to modify the perception of a particular part of the car: wheel arches, for instance, can be turned subtly outward at the top to catch the light, creating a highlight in what would otherwise be a dull surface. The same applies to the haunches, or exaggerated shoulders above the rear wheels, that characterize some Volvos and Jaguars; these produce focal points that reinforce the design's identity.

More interesting still are the so-called "light-catchers" that have been brought into play as high-riding crossovers become the style norm. These models often have deep sides that could appear bulky. To counter this, the lower portion of the door panels can be contoured so that it faces slightly upward and therefore reflects lighter tones, reducing the visual mass of the bodyside and thus the apparent height of the vehicle. Some makers, such as Renault, combine this with a "pinched in" effect created by a slight hollowing of the doors and lifting of the dark sill line. This helps the Clio and Scenic look lean, low, and lithe.

The influence of the paint finish is important here, too: blacked-out panels, for instance, can easily trick the eye. And though the confusing "flip-flop" colors beloved by customizers are thankfully out of the mainstream vogue, matte and semi-matte paints can lead to a totally different and potentially more interesting reading of the car's shape. What is reflected here is pure diffused light, rather than actual objects and shades.

THE ELEMENTS OF STYLE
LIGHTING, JEWELRY AND DECORATION

FORD'S SEQUENTIAL THUNDERBIRD

One of the most celebrated lighting effects is the sequential rear turn signal of the 1965 Ford Thunderbird. The signal illuminated three bulbs in sequence, giving the impression of gesturing in the direction the vehicle was turning.

BMW'S ANGEL EYES

Lighting supplier Hella struggled to find a customer for its luminous light rings until BMW snapped up the innovation for the 2000 5-Series. The model quickly became lauded as "angel eyes" and now almost every automaker provides a customized light signature.

STACKED TAILLIGHTS

Volvo's 850 wagon of the 1990s is credited with introducing full-height stacked taillights bordering its tailgate. But was it first? What about the German Ford Taunus 17M station wagon of the mid-'60s? Its taillights were in the roof, above the tailgate—and it pioneered rectangular headlights too.

Lights are among the most important graphic elements of a vehicle's identity. Some designers liken the headlights of a car to the eyes of a person, giving character and expression to the face. Indeed, some cars even appear to squint or frown; Morgan's original Aero 8 famously had a disturbing cross-eyed look until corrected on later versions.

Headlights are a sensitive topic too. BMW faced uproar when it dropped its characteristic four round light theme in favor of round lights behind flush glass covers in the E36 3-Series in 1990; others protested when Peugeot moved away from the slanting trapezoidal lamps that had characterized its models since the 1960s. More recently, models such as the Nissan Juke and Citroën C4 Picasso have dispensed with the two-eyed face in favor of a multi-light arrangement, while Alfa Romeo and Bugatti have used arrays of six or eight projectors for some time.

LED technology is unlocking amazing new possibilities for designers, something that's sure to stir up further furor. Already, most new cars display their brand message in carefully shaped frontal LED-light signatures; soon, always-on rear light signatures will become a key brand identifier too. LED forward lighting is compact and efficient and can be placed almost anywhere, so in place of the familiar eye-type headlamps we could soon be seeing lighting strips or bands.

Yet even with all these profound changes, lighting will continue to be an important marque signifier. Lights will always be one of the most visibly technical features of the car's exterior, and the scope for new shapes, signatures, and effects is unlimited. LED and fiber-optic illumination is already used extensively in interiors and it can easily be imagined as exterior decoration too, perhaps replacing conventional brightwork—with the ability to change color and intensity as well.

Another pointer to the future could be Mini's Next100 concept of 2016: it is able to back-project messages onto its door panels for passers-by to see, raising the possibility of cars that not only glow in the dark but also change color like a chameleon.

THE ELEMENTS OF STYLE
IDENTITY AND BRANDING

LOGO LOGIC
A distinctive logo is a key part of a corporate identity, and Mercedes-Benz, Audi, Renault, and Citroën score highly. Porsche's elongated script has the required sophistication too, but Citroën's premium offshoot DS takes the prize for sheer style.

GRAND GRILLES
Few would question the grandeur of the temple-like grille of a classic Rolls-Royce, nor the shock of Lexus's gaping scoop. But for sheer acreage of frontal real estate nothing beats North American SUVs with their immense, truck-like fronts.

ORNAMENTS, EMBLEMS, AND AFFECTATIONS
You used to be able to tell a Buick by the row of round portals on its hood or fenders, or a Wolseley by its illuminated grille badge. Most of these once-treasured devices have disappeared. Could modern brand-specific flourishes such as Jaguar's side air vents become tomorrow's totems?

What makes a Rolls-Royce so instantly identifiable? What makes a Mercedes so clearly a Mercedes, or gives a BMW its unmistakable image? Stature and badging apart, the most immediate answer in each case is a style of front grille that is unique to the marque and has been expertly evolved over the years to adapt to all the other trends in body design, aerodynamics, and even pedestrian safety.

Mercedes is consummately skilled in this regard: its identity is crystal clear, even though it now has at least three main types of grille and has made the major transition from a portrait profile in the 1960s to today's landscape format. It helps, of course, to have a brand emblem as distinctive as the large three-pointed star, and this holds true for most premium brands.

There is no clearer illustration of the significance of the grille as body decor than the struggle of aspiring premium marques such as Infiniti and Lexus for acceptance alongside the traditional premium players. Lexus has opted for an oversized "spindle" grille in its quest for at-a-glance recognizability. Infiniti is more subtle but has developed a distinctive C-pillar style which, at least to those in the know, signifies the brand in much the same way as BMW's much-cited "Hofmeister" kink, also in the rear pillar. Other brands have their own in-house conventions which have become brand signifiers: Land Rover's clamshell hood, Porsche's smoothly rounded forms, and even the sadly defunct Saab's distinct browed windscreen style.

For customers, however, it is the more visible features that count most, and recent years have seen an explosion in demand for outward personalization. Kicked off by the new-era Mini in 2001, options such as contrast-color roofs, stripes, decals, and multiple wheel designs have been an important priority for image-conscious buyers as well as commission-hungry sales staff. But ironically, once everyone has a similar set of individualization features, all cars will look the same again—and designers will face the challenge of taking brand identity in a new and more distinctive direction.

THE ELEMENTS OF STYLE
THE NAME OF THE GAME

CHANGE OF IDENTITY
Pity the poor Princess. British Leyland's wedge-shaped family car was launched with fanfare in 1975 as the Austin-Morris 18-22 Series. Less than a year later the lineup was rebranded Princess and in 1982 it was back to Austin again—as the Ambassador.

NAMES OR NUMBERS?
Premium marques tend to favor numerical or letter designations, such as C-Class, 7-Series, or Q70. This gives a technical impression and helps buyers understand the model hierarchy, important among blue-chip products.

MEANINGLESS, OR WRONG
To avoid trademark infringements, many companies invent new names: Aygo, Yaris, and Avensis are a few examples from Toyota. Perhaps better than manifestly inappropriate names: Mitsubishi Carisma (it had none), Suzuki Esteem and Daihatsu Applause (both singularly unmerited), and Toyota Urban Cruiser (it didn't) spring to mind.

A rose by any other name would smell as sweet, mused Juliet, famously, to her forbidden Montague lover, Romeo. Though Shakespeare may have been the most prominent of many literary greats to question the importance we place on the names we give people and objects, his talents might not have suited the motor industry that was to spring up many centuries after his heyday.

For manufacturers and promoters, the name given to a product is an intrinsic part of its image and, by implication, its design and the impression it projects. Names are imbued with auras, associations, and even innuendo, conjuring up pictures in the mind of the prospective buyer. Charger, Challenger, Boss, and Hellcat clearly stand at the opposite end of the spectrum to Leaf, Zoe, and Micra.

Numbers and letters, by contrast, are coldly neutral and only become as emotive as the product they are attached to. Just think of the contrasting auras of two cars called 3—the Mazda 3 and the BMW M3.

In the pre-globalization era, when each country was shielded by tariff walls and relied much more on its domestic producers than importers, manufacturers could get away with names attuned to local culture. So counties and cities were deemed suitable for the British, flower names for the Japanese, and palaces for French buyers. But international trade soon proved how many names failed to travel: Europeans sniggered about the Nissan Cedric, for instance, and how many Americans could pronounce Elysée?

Coming unstuck in translation was an even bigger risk: Mitsubishi's Pajero proved to be rude in Spanish and the Toyota MR2 sports car had to be renamed for the French market. Years ago, Rolls-Royce was said to have discovered at the last moment that Silver Mist did not translate well into German; Suzuki's Baleno is perilously close to "whale" in France, while Renault's Mégane means "glasses" in Japan. Nevertheless, Land Rover's determinedly Anglo-Saxon names do seem to travel well, reinforcing its British image.

But as top brands demonstrate, simple and clear is the best naming strategy. And if, like Mercedes-Benz, Audi, and Renault, you have a really strong logo, you don't even need to put your name on the car.

THE ELEMENTS OF STYLE
THE STYLE THAT LASTS

ITALY RULES

When it comes to beautiful cars, Italy rules. That may be something of a cliché, but examination does bear it out: not all the brand names may be Italian, but Italian design houses have shaped style-setting products for manufacturers on every continent.

TOP 100 BEAUTIES

Autocar magazine's 100 Most Beautiful Cars of All Time lists four BMWs, five Bentleys, five Bugattis, five Lamborghinis, six Porsches, seven Jaguars, seven Alfa Romeos, nine Aston Martins, and no fewer than *thirteen* Ferraris. The top spot in 2017 was taken by the Jaguar E-Type.

HOUSE OF HORRORS

The hall of infamy at the opposite end of the beauty scale has no shortage of candidates. At the head of the queue are the Mini Coupé, Pontiac Aztek, Mercedes GLK, Honda Clarity, Toyota Mirai, and two most recent generations of the European Honda Civic.

It's easy to be stylish if you're large, expensive, and above all, powerful and fast: a long, extravagant hood to conceal a mighty engine, wind-cheating curves to spell out speed, and finely executed detailing to hint at the classy engineering below. To stay stylish in the face of each passing fashion is less simple, however, and, as Coco Chanel pointed out, fashion is that which goes out of fashion.

And it is extreme fashion that can be the quickest to look dated: the elegant Lamborghini Miura of 1966 has lasted better than the wild Countach of 1974. A whole raft of Ferraris prove the point: they still look graceful, as do countless Jaguars from the XK120 to the F-Type; the Aston Martin DB4 is there, too, and today's BMW i8 may follow.

Smaller and more affordable models can add a dash of glamour to qualify, too. The shining example here is 1954's Alfa Romeo Giulietta Sprint, a true beauty even after six decades; Audi's original TT has lasted well, too, as has the A5 Sportback, and the innovative Range Rover Evoque still looks fresh.

But keeping that long-term shine is a lot tougher when your primary mission is not that of charm and seduction but the unglamorous task of everyday family transport. Names from the past that have worn exceptionally well are the Fiat 1100 of 1953; numerous Peugeots including the 404, 205, and 306; the Volvo Amazon; Renault 5; Lancia Delta; and Golfs Mk1 and Mk4.

The reborn Fiat 500 is still as chic as when it first surfaced in 2007, in contrast to the BMW-inspired new-age Mini which started strong and stylish in 2001 but which has descended further into clumsy caricature with each successive generation.

Classical good looks are another secret of long design life. The BMW 507 (1955) has an ageless beauty to it, as does the Citroën DS of the same year—as well as the Chrysler 300 and the Mercedes 230SL of the 1960s. But perhaps it is Porsche that has cracked it best of all with the 911, making changes between each generation so subtle that nothing looks passé.

GLOSSARY: CAR DESIGN

Beltline: The line directly underneath the side windows of the car, where the upper greenhouse meets the lower bodyside or shoulder. The position and inclination of the beltline affects the appearance and proportion of a car, as well as its character and stance.

Clamshell: A hood or trunk lid where the shut lines are extended outward to wrap round to the vertical side of the body. Typical of Land Rover and, in its heyday, Saab.

DLO: The expression derives from "Day Light Opening" and is used to describe the graphic shape of a car's side glass. The DLO is the strongest and most important graphic element of a car's design.

Fender (UK: Wing): Originally, covers directly shielding the wheels, bike-style, but now referring to the integrated bodywork areas above the wheel arches.

Firewall (UK: Bulkhead): The structural panel that separates the engine compartment from the passenger compartment and that provides sound and heat insulation.

Floating Roof: A roof design that appears to "float" as the pillars are blacked out or glazed over so they appear a similar tone to the glass; typical of recent Range Rovers.

Greenhouse/Glasshouse: The upper, glazed part of the passenger compartment, a key element in the vehicle's side profile.

Haunch: Where the shoulder of the car gently swells out to accentuate the muscularity of the rear wheel, often associate with the generic rear-wheel drive coupé form and, especially, Jaguars.

Hood (UK: Bonnet): The exterior body panel that covers the engine compartment of front-engined cars and that can usually be lifted or opened to provide access to the engine.

Light-catcher: A section of a panel specially shaped—usually indented—to face upward and provide a highlight by reflecting the light from above. Often employed to add interest to the lower bodyside panels, and typical of recent Renaults.

Lines: Feature lines, crease lines, character lines, and swage lines all contribute to the visual structure of a car and serve to create definition, add emphasis and interest, and direct the viewer's eye.

Overhang: Those parts of a car that project forward of the front wheels and extend rearward of the rear wheels. The relationship between overhang and wheelbase is critical in achieving an overall visual balance.

Pillars: The A-, B-, and C-pillars, counting from front to back, are important structural members; doors are hinged on to them, they support the roof and protect the occupants, and visually frame the windows. Pillars may be disguised graphically by being matte-blacked out, or by being wrapped by the side or rear glass. The shaping of the C-pillar can provide important brand identifiers, such a BMW's Hofmeister kink or Infiniti's reverse at the top of the pillar.

Plan Shape: The amount of curvature in bodysides, and particularly front and rear ends, as seen from above (in plan view). Clever use of plan shape visually pulls the corners back and enables designers to disguise excess overhangs front and rear.

Shoulder: The shoulder line basically runs the length of a car's upper bodyside where it folds over to meet the side windows and its nature will reflect the essential character of the car.

Stance: The attitude of the vehicle in relation to the road and to other vehicles. It is largely defined by the body-to-wheel and the overall vehicle-to-ground relationships, as well as the orientation of the beltline and the profile of the glasshouse.

Surface Language: The basic form language of the car that defines the design—it could be rounded, taut, angular or flowing, complex or simple.

Wheel Arch: The relationship of the wheel to the wheel arch is critical, and designers usually try to achieve as close a fit as possible: Audi is an expert here.

Wheelbase: The distance between the front and rear wheel centers, and a critical dimension in the quest for internal space efficiency and optimized accommodation.

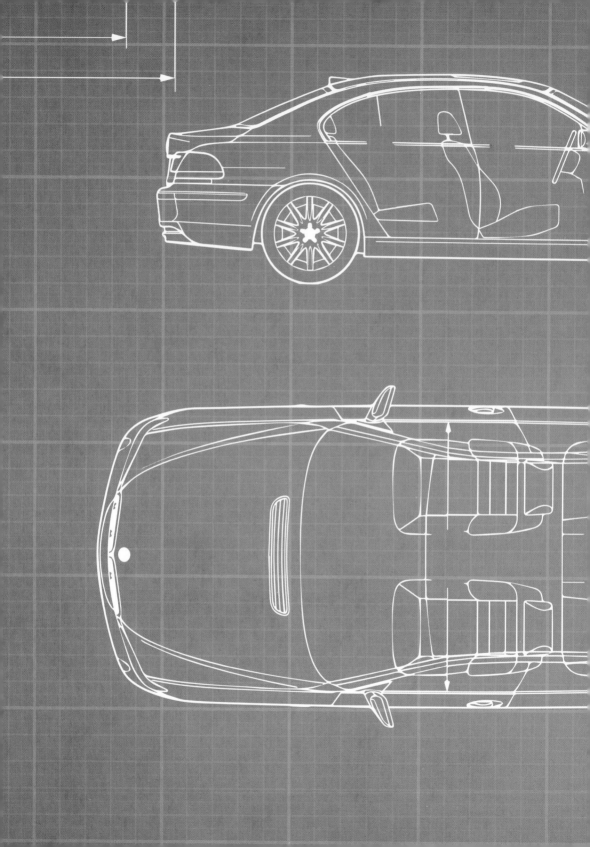

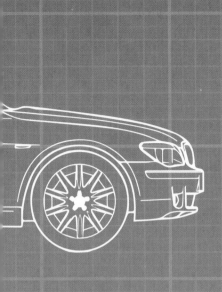

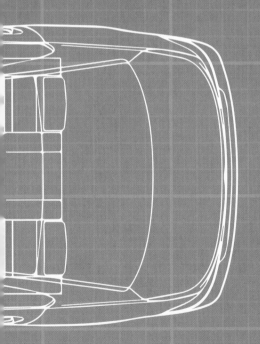

ENGINEERING, INTERIORS AND SPACE

INTRODUCTION
ENGINEERING, PACKAGING AND SPACE EFFICIENCY

In car design it is engineering that, even though hidden well away below shiny surfacing, determines almost everything. Dimensions laid down by chassis engineers provide the basic "hard points" that frame the passenger compartment and act as the starting point for body designers to shape the exterior proportions and style; the type and positioning of the engine will influence the shape of the front; the location of the fuel tank can affect the space available for passengers; and even suspension design matters—a complex rear-axle arrangement can intrude into trunk space.

The combination of all these things and how they all fit together are known as the package, and mark the start of the traditional design process. Packages come in all shapes and sizes but, with a smaller road footprint to play with, it is invariably compact cars that demand the most skillful arrangement of their components. For this reason, small cars have been responsible for most of the big advances in package efficiency—none more so than the Austin Mini, which in 1959 was the first to place its engine across the chassis at the front to allow fully four-fifths of its length to be devoted to passengers and luggage. Now, almost every car outside the luxury class employs this configuration.

Since the Mini's day the advances have been less spectacular, but no less significant. Mercedes bravely redrew compact-car architecture with its 1997 A-Class, laying the engine flat under the front occupants' feet so passengers could enjoy S-Class space within the length of a Polo; unfortunately, it survived only two generations in this form as it was too expensive to build. Honda's rethinking of the fuel tank allowed its Jazz/Fit supermini an innovative folding rear seat arrangement, and Toyota's iQ microcar even rethought the differential and heating pack in search of extra interior space.

Safety, too, has become an important essential in the packaging mix. Cars are now larger because of rules demanding adequate crush space front and rear for impact absorption, and the car's front has to be profiled to be as safe as possible in a pedestrian impact. Another reason why cars are becoming larger is that people themselves are getting taller and need more space—placing even tougher demands on the ingenuity of the package engineers.

ENGINEERING, INTERIORS AND SPACE
ARCHITECTURES, ENGINES AND DRIVELINES

TATRA'S V-8
It seemed a good idea in the mid-'50s when Tatra engineers decided on a V-8 engine for their large 603 limousine. But with the weighty air-cooled motor suspended behind the rear axle, handling was difficult. Even so, the 603 did chalk up some rally successes.

MOST UNUSUAL ENGINES
The ruthless ambition of one of the auto industry's greatest engineers, Ferdinand Piëch, means the VW group comes out on top. Porsche's race-winning flat 12, Bugatti's 18-cylinder concept and W16 Veyron, Bentley's W12, and VW's unusual VR6 and V5, as well as the original five-cylinder Audi, are all down to him.

HIDDEN ENGINES
Cunningly concealed engines have included the central underfloor motor of the Toyota Previa minivan, the air-cooled flat four in the rear of the VW Microbus, and the tiny three-cylinder squirreled away ahead of the rear axle in Mitsubishi's "i" microcar.

We have just seen how the package of a vehicle is wrapped around a set of fixed "hard points" that are traditionally dictated by the physical shape of the engine and whether it drives the front, rear, or all four wheels. More recently, however, manufacturers have been moving toward a modular approach, with sets of standardized parts that all fit together and dramatically improve manufacturing efficiency and allow many more permutations of length, width, body style, engine type, and drive arrangement.

Even in the most sophisticated of these architectural grand plans some parameters remain fixed, typically the dash-to-axle dimension that governs how much space there is for the engine assembly behind the front bumper. Within these confines, the vehicle can accept a wide variety of engines, in VW's case spanning gasoline, diesel, hybrid, and pure electric. Nevertheless, the vast majority of the world's car population has standardized on the transverse engine, front-wheel drive layout, generally with four cylinders but sometimes three, two, five, or even six. A rare version of the 1980s Lancia Thema had a Ferrari V-8 up front.

Like the once-universal front engine, rear-wheel drive is now a relative rarity, confining itself to sporty or luxurious cars. Engine choices here can be anything from four-cylinder inline to extravagant V-12s, the latter posing something of a packaging challenge—just witness the complexity under the hood of an Aston Martin or Ferrari. Bugatti's Veyron and its Chiron successor boast centrally-mounted engines with no fewer than 16 cylinders, and some front-engined sports cars even position the gearbox in unit with the rear axle so as to improve handling balance.

In terms of engines themselves, flat or horizontally opposed types are rare but have always been the most appealing to vehicle packagers: they are compact and well-balanced and keep the center of gravity low for good handling. Remember the flat four that disappeared under the rear floor of the VW Type 3 Variant in the 1960s, or Citroën's GS, where it sat ahead of the front wheels under a low nose? Or the Porsche Boxster's invisible flat six, mounted just ahead of the rear axle?

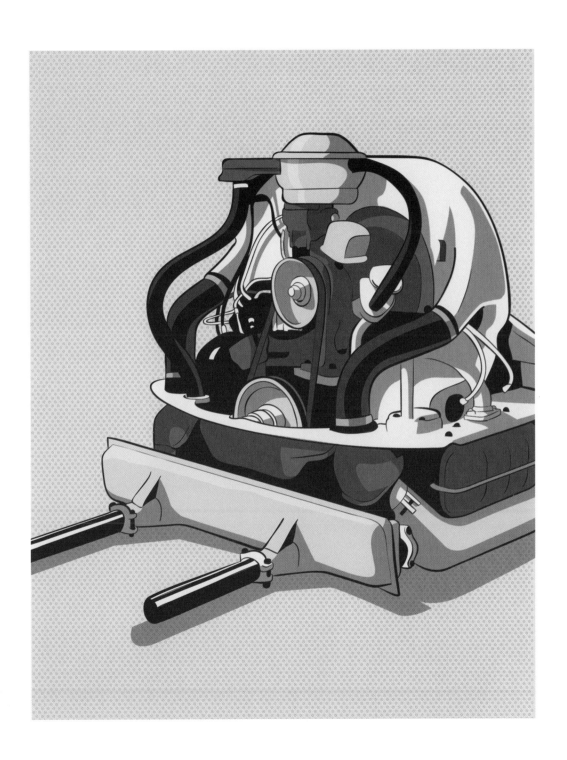

ENGINEERING, INTERIORS AND SPACE
THE REVOLUTION AHEAD

THE INCREDIBLE SHRINKING FUEL CELL

Fuel cells, which produce electricity from hydrogen and oxygen, emerged from the US space program in the 1960s, prompting GM to develop its 1966 Electrovan prototype. The bulky apparatus in the back turned the large six-person minivan into a cozy two-seater. Today's fuel cell stacks fit neatly under a standard hood.

GM'S HY-WIRE CONCEPT

To demonstrate the potential of its fuel cell sandwich chassis, GM turned to Italian design house Bertone in 2002 to add a stylish body and radical drive-by-wire control system. The prototype's spaciousness was remarkable, with the flat floor completely unobstructed from front panel to tail panel.

SIMPLE PROMISE

What could be simpler than an e-wheel? Each wheel has its own electric motor, transmission, suspension, and brakes and, for the front of the vehicle, steering. Just add cables, battery, and a control system.

With the coming of new forms of propulsion, such as battery power, fuel cells, or even compressed gas energy, designers are likely to enjoy far greater freedoms in how they configure their vehicles, in turn opening up entirely fresh possibilities for exterior shapes and arrangements for passengers and luggage within the interior.

One of the first studies to bring this home was GM's Autonomy concept of 2000. Forget the spiky, racer-like body, the real interest was in the thin, skateboard-like chassis, which contained all the hydrogen fuel cell equipment, control systems, and drive motors. This opened up the industry's eyes to the potential of electric power—whether through batteries or fuel cells—to completely rewrite the rules on vehicle packaging.

Electric motors are much smaller than combustion units and require less cooling and fewer ancillaries; engineers are even developing electric motors built into wheel hubs, so all that the vehicle theoretically needs are a battery, a control system, and cables to the two—or four—wheel motors. French specialist Venturi demonstrated its Volage sports coupé concept in 2008, using a Michelin e-wheel at each corner. For sports aficionados this approach is likely to open up new frontiers in handling dynamics, but perhaps more importantly it gives designers wholly new freedoms in how they package the vehicle. No longer do the principal components need to be mechanically connected to one another with bulky shafts and couplings. Now, items such as the battery and electronic modules can be placed wherever the designer feels is most advantageous.

To date, however, no production model has yet provided a convincing demonstration of the space-saving potential of electric power, though the luxurious Tesla Model S manages to provide five adult seats, two pop-up child seats, and both a front and a rear trunk—all within the footprint of a five-seater Mercedes E-Class. Yet as a taster of what is to come, models from Volvo, Toyota, Peugeot, BMW, and Nissan now feature electrically powered axles as a means of adding all-wheel drive and hybrid operation—with barely any adjustments needed to the overall vehicle architecture.

ENGINEERING, INTERIORS AND SPACE
INTERIOR ARCHITECTURE

SIT UP STRAIGHT

An upright seating position may make a car taller but it helps provide more passenger space within a given length. Witness the 1983 Fiat Uno, inspired by Giugiaro's high-riding Megagamma concept, and the original Ford Focus.

DOORS: DULL TO DRAMATIC

So-called "suicide" rear doors that hinge at the back are too tricky to engineer for most normal cars. There is more variety in wagon and hatchback access, with single or split tailgates, double doors, or a single side hinge. But if it is an exotic supercar the doors must be theatrical, opening upward to please the crowds.

MINI REINVENTS THE DOOR

Since its revival by BMW in the 2000s, Mini has made unusual door design a brand highlight, the 2007 Clubman mini-wagon kicking off with a curious half-length extra "Club" door on the right-hand side, as well as twinned side-hinged barn doors to the rear load area.

The key starting point in defining the basic package of a vehicle, as described in the preceding sections, is the number of seats: a two-seater coupé will clearly have very different proportions than a seven-seater minivan. How high the seating position is—the so-called H-point, or hip point—is important, too: in the sports car this will be close to the ground, while in family vehicles the current vogue is to sit high up, crossover style, for a sense of security and a good all-around view.

Sitting more upright generally allows a better use of vehicle length, while the sports car driver with legs extended almost horizontally is more wasteful. But for each of these, designers will have in mind the size of occupant and whether to cater for truly large individuals in terms of headroom, shoulder width, and seat adjustment.

There is some flexibility in the arrangement of seating, especially where it really counts—in small cars. The Toyota iQ's 3+1 layout squeezed four people into its tiny cabin by hollowing out the dashboard and having the front passenger sit farther forward than the driver to allow space in the rear. Fiat's Multipla and Honda's FR-V had 3+3 seating; other people carriers choose three rows to house 2+3+2. Many hatchbacks have sliding rear seats so that families can strike their preferred balance between back-seat space and luggage volume; how seats fold, recline, and slide has to be engineered in, as does the always-tricky access to the third row in a minivan.

In establishing the basic seating positions, designers will have taken into account anthropometric data and will generally try to accommodate drivers from a fifth percentile female of 150 cm (59 inches) and 40 kg (88.1 pounds) to a 188 cm (6 feet 2 inches) 95th percentile male weighing over 120 kg (264.5 pounds); all these people must be able to have adequate sightlines, reach the main controls, and have suitable seatbelt paths and airbag protection. Smaller passengers, children, and infants in child seats need to be taken into account too.

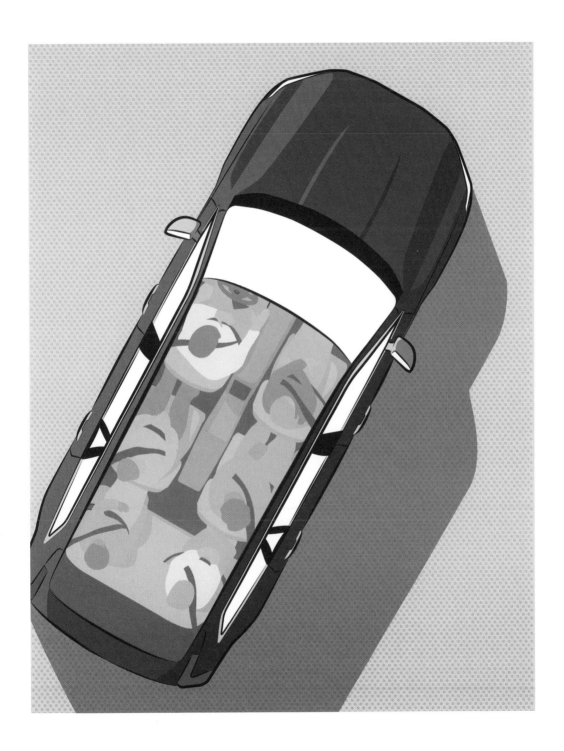

INSTRUMENTS AND DISPLAYS

DISPLAYS OF ECCENTRICITY

Citroën has always relished doing things differently. Its GS and CX of the 1970s used spinning drums behind glowing lenses to display speed and engine rpm—not great for clarity or at-a-glance legibility.

MCLAREN'S SHAPE-SHIFTER

For its 2017 supercar, the 720, McLaren rethought the information display and came up with a novel foldaway instrument panel. For road use a full-size screen gives standard driver information, but in track mode the screen swivels down to reveal a slender linear engine revs display, just as in a racing car.

INFORMATION OVERLOAD

Seamless integration of smartphone operation has become an essential selling point. With most cars now including a central display screen, familiar iPhone-style icons have come to the car designer's rescue to provide simple and safe shortcuts to music, communication, and navigation.

When it comes to technology directly visible to the customer, this is where the biggest change of all is occurring. Just as in other product areas where the electronic display of information has become dominant, so in car design the screen—and touchscreen—are steadily taking over from conventional mechanical and electrical instruments and controls.

Electronic displays offer numerous advantages. They are more compact, reliable and, versatile; they are simpler to install and they can switch between modes to suit the whim of the driver or the type of conditions being encountered. No one needs a rev-counter in the middle of city traffic, so the unit can switch over to mapping and navigation; equally, for top-end sports cars a "track" mode can highlight engine and chassis parameters of importance to drivers chasing a quicker lap time.

There were many baby steps on the route to today's digitized dashboards. Early electronic instruments in the 1980s had garish graphics of dubious accuracy, but it was Lexus in 1989 that produced the first solid-state display truly acceptable to premium customers. Lexus wisely decided to stay with analogue needles for most functions, something that has remained the norm in the upper segments, even though those needles and scales are generated electronically. Again, the quality of the graphics is the crucial factor and Jaguar was one of the first to look genuinely authentic. Digital read-outs for speed are well accepted in city cars and family vans where their clarity and compactness are valued. Head-up displays project information onto the windshield and are now becoming available on a wide range of models.

The growing amount of driver-facing information will be a major issue for designers in the years to come, especially the need to avoid the distractions of phone calls, messages, and navigation. Two contrasting solutions are already available: Tesla's giant center-mounted portrait touchscreen that controls almost everything on the vehicle, and Audi's brilliant reconfigurable dashboard—essentially a dynamic moving map set as a background to the standard instrument pack, but with the driver able to enlarge, reduce, or select from any number of display options.

ENGINEERING, INTERIORS AND SPACE
DRIVING CONTROLS

Touch, and tactile sensation in general, are too often overlooked in the makeup of a vehicle, but given that much of the enjoyment of driving is felt through the seat and the steering, these qualities deserve careful consideration.

The tactile experience begins with a tiny item, the key. Does this feel light and flimsy in the hand, or solid and high in quality? Is that first impression good or bad? Next, the exterior door handles, again a quality signal if they are firm and substantial; and the weight and sound of the door as it is opened and closed—all deliver messages about integrity and quality.

It could be argued, quite fairly, that those inputs are peripheral to the actual activity of driving. What counts more, for most driving fans, are the actual dynamic experience and the feel of the chassis, the engine, and the controls—especially the steering, the gear change, the pedals, and even the minor buttons and switches. It's not enough for the steering wheel to be at the right height and stretch; it must be the right diameter, the rim must be just fat and firm enough and have a good surface texture for rapid swerves and maneuvers. All these qualities will come naturally to a company that knows and understands, like Porsche.

Gearshifts, whether manual or automatic, are another sensitive area. A manual lever needs short, precise movements that feel mechanically satisfying but not too light and remote; an automatic selector requires clear detents so that it can be worked intuitively without looking, and the lever itself needs to feel well engineered rather than cheap and insubstantial. Even the sounds made by these control movements contribute to the quality picture: there is nothing worse than the grating ratchet rasp emitted by a US-style foot-operated park brake as it is applied.

Switches are a tactile element too: just notice how visitors in auto shows or car dealerships prod and poke at every button and knob. Control stalks also play strongly into the quality impression, with any hint of flimsiness or plasticity earning an immediate negative—and this now applies to lower-priced cars as much as premium models.

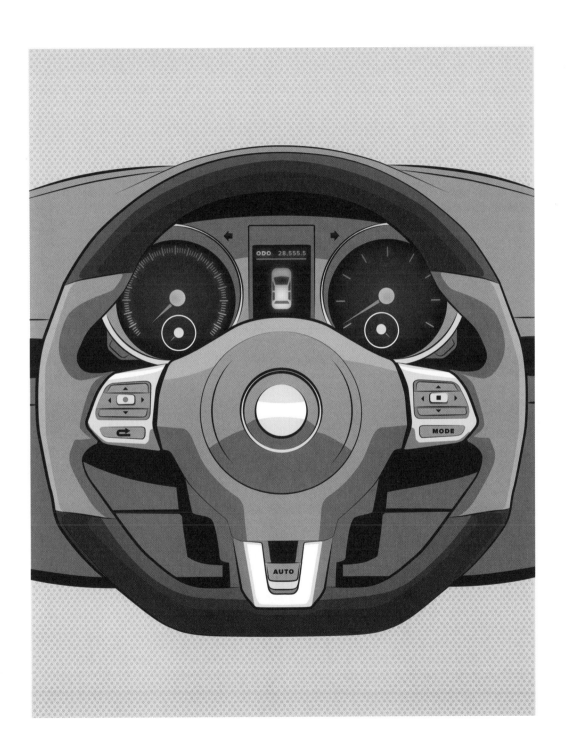

AUDI DESIGN

Volkswagen's premium brand Audi operates what is by common consent one of the slickest and most highly regarded design operations in the car business. From shaky beginnings in the 1960s with stretched small DKWs, Audi found its feet with the first Audi 100 sedan in 1968 and was applauded for its elegant Italianate 100 Coupé in 1970. But it was the super-aerodynamic third-generation model in 1981 that brought the brand universal recognition—even though to today's eyes the streamliner looks somewhat over-bodied and indulgent.

The follow-up was more conservative and restrained, but it marked the beginning of the precise and restrained form language for which Audi would become renowned;

the smaller third-generation Audi 80, with a notably neater and more compact look, had already set a new high-water mark in interior design with its smooth fascia flowing seamlessly into the front door pulls. The fit and finish and the clean simplicity set it apart from its fussier mid-market competitors, and the red-glowing instruments were a classy touch.

The TT sports car a decade later saw Audi take another bold step in interior design. Milled metal surrounded the dashboard air vents, giving a reassuringly engineered feel, and diagonal bright metal struts braced between the dashboard's center stack and the lower console carrying the gearshifter—also technical in its appearance, with a metal bezel—and parking brake. The overall impression was that of innovation and sporty integrity, important given that the model was based on components from the rather less sporty VW Golf.

When BMW launched its E65 7-Series in 2001, the shock of its bulky styling was nothing compared with the storm over its iDrive controller operating almost all its secondary functions: few could fathom the intimidating system, and BMW gradually simplified it. But Audi's take on iDrive, MMI (for multi-media interface) appeared the following year in the A8 and proved much easier, more intuitive, and more effective.

While the architecture and impeccable quality of Audi's interiors has remained consistent for a decade or more, the so-called virtual cockpit, or reconfigurable dashboard, marked a major step forward when first seen on the third-generation TT in 2015. As outlined briefly earlier in this chapter, in place of the normal instrument cluster is a solid-state screen on which can be displayed everything from high-resolution navigation mapping to the MMI function interface and even classical round instruments, as well as combinations of all these. The beauty of the system, especially for a compact sports car such as the TT, is that it enables designers to dispense with the bulky central navigation screen and free up valuable space on the dashboard.

And now, with the fifth-generation A4 and, most recently of all, the new 2018 A8, Audi has further refined the clean elegance of its interiors. It may not be the fanciest or the flashiest, but it is the most consistent and, many would contend, the best there is.

ENGINEERING, INTERIORS AND SPACE
COLOR AND TEXTURE

GO MATTE

Lost in an ocean of shiny metallics, pearlescents, and high-luster gloss? New matte and semi-matte finishes could be the answer. BMW's Frozen Metallic series and Mercedes's Magno paint lines throw an intriguing new light on body surfacing and work especially well on sportier models in grays and gentle silvers.

PLASTIC PROGRESS

Poor-quality interior plastics used to be a perennial complaint of all but the poshest wood-clad cars. Now, thanks to clever "slush" molding techniques pioneered by VW in the 1990s, the luxury feel of a smartly textured soft-feel fascia molding is a near-universal privilege.

MATERIAL CHALLENGE

Interior fabrics and materials undergo tests so agressive that only a few make the grade.

Once the bare-metal body has been painted, assembled, and fitted out, what was once the designer's sketch has become a physical vehicle that can be touched, smelled, and experienced. This is a transformation that's largely down to the "soft" values added by what the auto industry refers to as color and trim. These are qualities that are vital for most buyers, and major carmakers employ many thousands in special design sections devoted to creating the desired ambience around—and inside—the product and the brand itself.

Color is about much more than just paint; color palettes come and go on a regular cycle as fashions change. The current favorites are on the silver/gray/black spectrum, with some pastel shades for smaller, retro-style cars such as the Fiat 500. Paint technology is advancing too, with attractive matte finishes available in addition to the standard solids, metallic, and occasional pearlescent. Exteriors have seen a return of chrome embellishments, along with a new material—charcoal-colored carbon fiber inserts to give a technical look to aerodynamic add-ons.

The big change in the past decade has been the rise of personalization. Small-car buyers in particular are able to specify anything from a contrast-color roof to racing stripes, and color-keyed interior styling accents and motifs woven into seat fabrics. For pricier models the changes are slower in coming and the conventional signifiers of luxury and sophistication still hold—polished wood on the dashboard and door cards, full or partial leather for the seats, and metal or pseudo carbon trim inserts to switch the ambience from gentleman's club to sporty technical.

Overall, it has been the cheaper and more youthful cars, along with avant-garde "early adopter" models such as the BMW i3, that have led the charge against interior design uniformity. The BMW's pale wood dashboard option gives a real architectural feel, like a fashionable urban loft residence; other producers have toyed with exotic upholstery materials such as silk and sports fabrics.

ENGINEERING, INTERIORS AND SPACE
THE DETAILS THAT COUNT

GM NEVER UNDERSTOOD SAAB
Despite controlling the Swedish carmaker from 1990 to 2009, GM never really understood Saab, believing that Saab character could be infused into an outdated Opel Ascona solely by the quirky central positioning of the key.

VW VENTS ITS FRUSTRATIONS
Former VW patriarch Ferdinand Piëch hated air vents in cars, so for the Phaeton, he decreed the air outlets should be invisible when not in use. This meant individual wood veneer strips applied to each of the dozens of motor-controlled vanes.

CITROEN'S SINGULAR APPROACH
To the legions of fans in Citroën's heyday, the single-spoke steering wheel was not a detail but an article of faith. It appeared in various forms for four-plus decades until cost pressures forced standardization—and a multi-spoke approach. Now it may pop up on the occasional concept.

We mentioned earlier the importance of seemingly small details such as keys to the perception of a vehicle. As well as providing a first tactile impression of the car, the key is something that always stays with the owner: it's like a phone, or a handbag, placed on the bar or on a restaurant table. As such, it's a proxy for the vehicle's quality and style—so the canny designer will ensure it is an object of pride and fascination too. New-generation keys can remote check vehicle status or even back the car into a parking space—again a source of fascination and intrigue that adds to the collective brand magic.

Of course, by any objective measure, these details are nowhere near as important as, ride comfort, fuel economy, or crash safety. Except that customers like details, perhaps disproportionately—which means that automakers actively seek out such novelties for buyers to latch onto. Ford marketing coined the trite phrase "surprise and delight" for features that do just that—cause spontaneous expressions of approval. One such, albeit from rival carmaker Volkswagen, caused a stir when presented on a new Golf in the 1990s. The interior grab handles, instead of snapping back noisily when released, returned to their rest positions silently in a slow and smoothly damped movement. Competitors scratched their heads in disbelief at how VW had managed to engineer such a sophisticated and costly solution, but for VW it worked a treat, helping cement a near-premium image for the model.

Earlier, VW had scored another minor but symbolic coup by using the same classy-looking remote as Mercedes in its then-current S-Class, while going back even further, Renault was first in the volume market with its "plip" remote locking—a sure attention getter in the company parking lot—and small Rovers impressed with a courtesy light fade-out normally associated with big Jaguars.

Where to draw the line between transient gimmicks, valid brand identifiers, and worthwhile, loyalty-enhancing innovations is a tricky question. VW's once-trendy blue instrument backlighting has been dropped—and would a Ford Mustang be any less Mustang without its triple rectangular taillights?

GLOSSARY: ENGINEERING, PACKAGING AND SPACE

Architecture: As a general term, the way in which a design is laid out as regards positioning of the various major components. More recently, a predefined set of vehicle components ("modules"), including chassis structure, suspension, brakes, and electrics, that can be standardized and built in large numbers to minimize unit costs.

Carbon Fiber: Light and very strong structural material used principally in racing cars. Small sections of carbon fiber lookalike material are becoming fashionable as decorative elements for interior and exterior of standard cars.

Crumple Zones or Crush Space: Front and rear sections of the vehicle designed to collapse in a collision, helping absorb crash energy to protect passengers.

Differential: Geared device fitted inside an axle to allow left and right wheels to turn at different speeds while negotiating a turn.

Drive-by-Wire: A means of operating controls such as steering and brakes using electronic commands rather than direct mechanical or hydraulic linkages.

E-Wheel: A roadwheel assembly comprising an electric motor and sometimes steering, braking, and suspension.

Electric Axle: Axle assembly, usually for the rear, containing its own electric motor, gearing, and differential. Enables front-drive vehicles to add all-wheel drive and hybrid operation without major structural revision.

Engine Positioning: The orientation and location of the engine within the chassis. The most popular is across the chassis (transverse) at the front, giving simple drive to the front wheels. Pre-1960s, the norm was a front engine set inline with the chassis, driving the rear wheels. Engines can also be centrally mounted, behind the passengers (mid-engined) or behind the rear axle, as in the Porsche 911.

Engine Types: Engines can have their cylinders inline, horizontally opposed (flat), or in a V configuration, with anything from two to eight cylinders or, exceptionally, ten, twelve or sixteen.

Grab Handles: Handles within the cabin to help passengers brace against cornering forces and assist with entry and exit. Also known as assist handles.

Hard Points: Fixed points in the engine/chassis assembly, around which the superstructure and body design are constructed.

Heating Pack: HVAC, or heating, ventilation, and air conditioning. A substantial module, usually located in the dashboard space or within a double floor.

H-Point or Hip Point: How high above road level the driver sits within the vehicle.

Hydrogen Fuel Cell: A propulsion system that mixes hydrogen fuel and oxygen from the atmosphere to produce electric current and power a traction motor. The fuel cell has no moving parts and its only emission is water.

Packaging: The art and skill of fitting people, luggage, and mechanical elements within a given vehicle envelope. A well-packaged design will keep the engine compact and devote proportionally more space to accommodation.

Parking Brake: Also known as hand brake in Europe, where it is generally operated by a pull-up lever between the front seats. North American vehicles often have foot-operated parking brakes but the most recent trend is toward electrically actuated systems.

Pedestrian Impact: Designers must shape vehicle fronts so as to cause minimum injury in an impact with a pedestrian.

Reconfigurable Dashboard: Electronic instrument display that can be switched between modes to show different types of information such as engine parameters, chassis settings, or navigation prompts.

"Slush" Molding: A molding technique used extensively for large interior components such as single-piece dashboard modules. The technique allows a high-quality surface finish backed by soft-feel foam whose density can be graduated across the component.

Solid-state Display: Instrument unit that uses electronics rather than physical needles and scales to display information.

"Suicide" Door: Magazine term for passenger door hinged at the rear, seen principally on Rolls-Royce models and certain minivans.

Tailgate: Rear door system giving access to the cargo area on hatchbacks, station wagons, and vans. Commonly top-hinged and in one piece, but can also have a separate downward-opening rear section (Range Rover). Mini Clubman has twin side-hinged doors and some SUVs a single larger side-hinged door carrying the spare wheel.

Transaxle: Placing of the gearbox and sometimes also the clutch in unit with the rear axle. Used in certain high-performance sports cars.

Transmission Tunnel: A raised channel pressed into the floorpan to carry the propeller shaft to the rear axle on rear-drive vehicles.

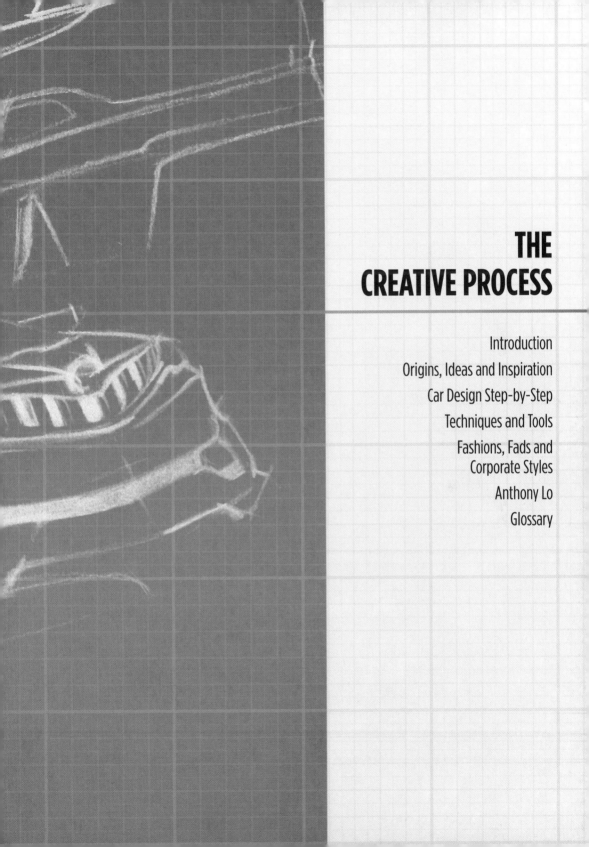

THE
CREATIVE PROCESS

INTRODUCTION
DESIGNING, STYLING AND CREATING

Never make the mistake of referring to a designer as a stylist. The term "stylist," for many, trivializes the art of creating a visual identity. "Stylist" carries connotations of a bygone era when autocratic engineers ruled the roost and would grudgingly pass their completed creations down to the styling department for a few finishing touches.

A modern designer, on the other hand, is responsible for much more than mere surface decoration. Designers no longer simply determine what the product will look like: they come up with new ideas for how it can be used too. They will be involved from the very first stages of a program, which could be the planning of new products, the drawing up of the design brief, through all the familiar creative design stages to marketing and product launch.

The art of car design emerged in the early 20th century as wealthy customers briefed skilled craftsmen on how they would like their car chassis clothed and bodied. The ideas, sketched on paper, would become reality in the coachbuilder's workshop as sheet metal, leather, and often exquisite wood were brought together to make a bespoke hand-built vehicle. It was only with the coming of factories and large metal presses that designs had to be translated into precise engineering drawings so that parts could be standardized and fit together accurately on the assembly line.

The designer's art of bringing ideas into three-dimensional reality gradually expanded over the years until all areas of the vehicle—including the interior—were thought through and designed as a coherent whole. The big revolution came as computer-based digital design began to make inroads into the conventional paper draftsmanship techniques. Initially, digital equipment was employed to accurately measure the finished clay model to ensure the utmost precision in the production tooling; later, digital creative processes began to offer a parallel path to the familiar paper sketching and quarter-scale clay model techniques. A major step was the digitization of the design at an early stage, allowing the model or component to be viewed on screen in three dimensions and from every angle. Currently, virtual reality (VR) is making inroads too, allowing the design to be placed into a real or imagined setting, viewed in motion in traffic, and even explored life-size, inside and out.

The realism of such viewing systems is breathtaking, and it could now be theoretically possible for a vehicle to be designed, engineered, and refined entirely in a virtual environment without a single paper sketch, clay model, or physical prototype. But even the strongest supporter of digital design would counsel caution in this: there is no substitute for viewing—and touching—a full-sized model, assessing it from every angle and in every type of light. And, as designers repeatedly insist, digital modelling cannot resolve the very trickiest details—so hands-on techniques still have the final say.

THE CREATIVE PROCESS
ORIGINS, IDEAS AND INSPIRATIONS

Precisely where new ideas spring from is often a mystery. In the car business they can be born out of necessity, perhaps to update an aging product or respond to a competitor's action. New design thoughts can be forced by a change in legislation, such as rules requiring front-end shapes to be less harmful to pedestrians, or they could respond to technical opportunities, such as the advance of electric propulsion.

Equally, societal developments such as the rise of urban living and the demand for constant smartphone connectivity can inspire fresh design thinking, as with the Nissan Juke, while the early identification of demographic changes and help designers create new segments to gain first-mover advantage—as Renault did in the 1990s with the Scenic family minivan and BMW and Fiat with the Mini and 500, respectively, both of which responded to affluent downsizers' desire for a premium small car.

Rarer is genuine blue-sky thinking, where opportunities are spotted for a wholly new type of product. The prime example is the 1964 Ford Mustang, the first car that was personal rather than purely practical. Another instant hit was the 1997 Toyota RAV-4, which made SUV driving both accessible and fun. Those that dared to be different but didn't quite hit the mark include the Smart two-seater of 2001 and the 2012 Renault Twizy.

The overwhelming majority of design projects, however, are scheduled as part of a comprehensive product plan and their main parameters are known well in advance. Here, the designers need to be just as imaginative: it may be that they can apply some ideas already in the pipeline, market research may have some useful pointers, and the values of the brand will of course need to be included.

All the while, the designers will have been keeping their eyes and ears wide open for broader trends, attending everything from fashion shows to furniture and product design fairs. And it is this extra perceptiveness in interpreting all these cultural triggers that enables a good designer to stay ahead of the game.

THE CREATIVE PROCESS
CAR DESIGN STEP-BY-STEP

3D PRINTING AND VIRTUAL REALITY

Both of these new techniques can enrich the vehicle development process. 3D printing can generate small prototype parts in the studio, while VR is an important presentational aid as well as useful in interior design.

CONCEPTS

Concept cars can go much faster from initial idea to final product ready for auto show glory. There are no tedious regulations to comply with, powertrain envelopes can be ignored, and tire-to-bodywork clearances need not spoil the party.

RACE TO MARKET

In the late 1990s when Japanese automakers replaced their models on a three- or four-year cycle, time to market was an important competitive factor. Among the fastest were some Japan-market Mazda models, which claimed a remarkable 18–24 months between design freeze and showroom launch.

Few car designs actually originate in the studio: almost all begin their lives in product planning, where a market requirement will be identified, engineering elements decided, and target dimensions and price point established to generate an initial specification that is then communicated to the studio.

Step 1. Concept Generation: Designers submit ideas reflecting the chosen themes and targets. Ideas from different sketches are often combined until a handful of themes remain.

Step 2. Shortlisting: The chosen themes are taken forward in electronic format, digitized in order to allow 360-degree viewing and initial sharing with key management personnel. Some designs may be eliminated at this stage, or new alternatives may be added.

Step 3. Final Shortlisting and Modelling: The favored designs will be taken further into more detailed 3D modelling or even rendered as clay models, generally ¼ scale, to allow presentation in a variety of lighting and background conditions.

Step 4. Interior: Often the design process for the interior will have begun running in parallel with the initial exterior design modelling, as soon as the key hard points have been fixed. Electronics specialists begin exploring architectures for the dashboard and display systems, and color and trim designers apply their expertise to possible textural palettes. Generally, fewer interior design themes are pursued.

Step 5. Full-Scale Model Build: With the final exterior and interior themes chosen, 3D models are milled in clay for initial approval by senior management and possible further refinement by designers. Then the clay model is accurately re-scanned to create detailed data to be used by engineering and, later, manufacturing.

Step 6. Final Approval: Procedures vary from company to company, but in most cases once management has approved the final design and engineering has verified that no hard-point changes are necessary, a styling freeze is imposed and the prototype testing phase is stepped up. By now the model is between two and three years away from commercial launch.

THE CREATIVE PROCESS
TECHNIQUES AND TOOLS

THE VIRTUAL INTERIOR
Virtual reality programs are especially useful in interior design. Once the main volumes of the dashboard, the door panels, and the switch-gear have been fixed, VR can be used to determine whether everything is reachable.

DIGITAL DESIGN TOOLS
The most important design software for the automotive industry comes from Autodesk's Alias suite. It covers everything from concept development through final class-A surfacing, while Showcase handles 3D presentation and rendering and SketchBook Pro is a cut-down sketching and drawing tool.

BMW'S MIXED REALITY
Claiming to be able to evaluate more design options and progress designs further before prototypes are built, BMW is combining 3D printed prototype components with a virtual reality environment developed using Unreal Engine's computer games technology.

The world of music has been rocked by digital technology, and songs created on a bedroom laptop can become worldwide hits. It would be perfectly possible for someone with the right software to produce a professional-looking car design on a laptop, too. But that's not how it works in practice. Who better to explain than Antony Lo, vice president of exterior design at Renault's design studio near Paris.

"There's a myth going around that designers both inside and outside the car industry only work with computers," he said. "That's not true. Most of them sketch on paper to start with. A computer with Photoshop or Illustrator is good for coloring, as I call it, but for generating ideas it's best to work with a tool you're comfortable with because you'll be sketching sometimes in a coffee shop, in meetings, in your office. Sketching is where most of the ideas come from. But when you want to make it into a presentation, say on a 5x4-meter screen, it needs to be more glamorous and we want to add in highlights and reflections. This is where digital tools can save us a lot of time.

"The other tools studios use for 3D design are Alias Autodesk. Designers should all know how to manipulate a [digital] model, put in some lines, and create the framework for the professional digital modeler to finish the model. It's not the job of the designer to build all the surfaces and so on. It's very complex and time-consuming if you are not an expert.

"The digital data is used to mill the first clay model, and then we work on this for a few months by hand in a parallel process. We do a lot of refinement on the clay model because at the end of the day there are limits to the digital tools; sometimes you can see things that look good on the screen but don't look so good in reality. So we work and scan every two weeks to rebuild the digital data as we have to constantly share data with engineering so they can keep track of the model and tell us for instance whether the door can open or the cooling is sufficient."

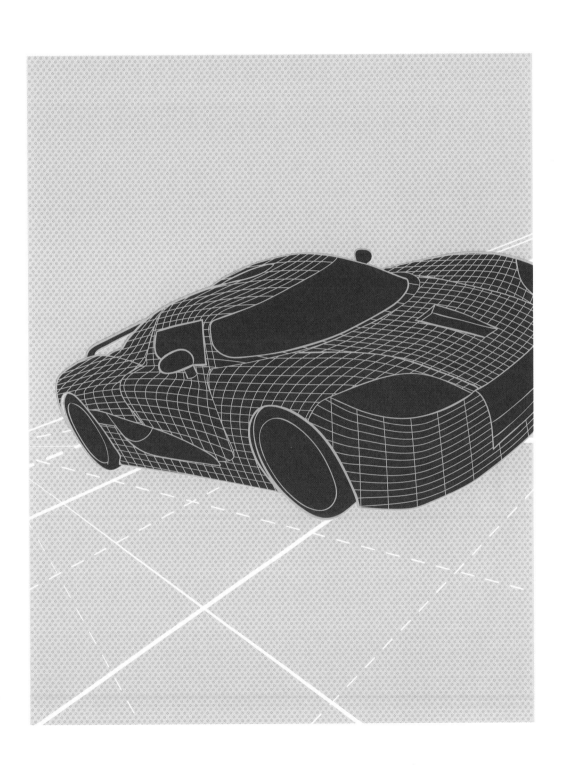

THE CREATIVE PROCESS
FASHIONS, FADS AND CORPORATE STYLES

AUDI'S SINGULAR GRILLE
The style did not have a name, but Audi's full-depth grille, first seen in 2003, introduced a clear vertical graphic into the vehicle's frontal face and was widely emulated. Look no further than Lexus, with its dramatic ground-hugging "spindle" grille.

VOLVO REINVENTED
Volvos were resolutely boxy, brick-shaped, and unstylish until the early 1990s when design director Peter Horbury set about changing consumer perceptions. The design language softened, curves and highlights appeared, and a quarter century later Volvo is a stylishly credible, near-premium global player.

RENAULT'S LIFECYCLE
Love, exploration, family, work, play, and, finally, wisdom: these are the six stages of life, symbolized as the petals on a flower that Renault has been using since 2010 to bring focus to its launches of concept and production cars.

Art and Science, Sensual Purity, and Athletic Elegance—just a sample of the dozens of corporate design philosophies to emerge in recent years. These PR-generated labels may seem meaningless, but they do provide a rallying call to designers—often under a new-broom studio head—and show consumers how seriously the company regards design.

The trend toward naming new design languages intensified after 2000, when Cadillac proclaimed its products would follow a new "Art and Science" philosophy—bold, square, and sharp—and Lexus and Toyota sought to shake off their bland images in 2003 with L-Finesse and Vibrant Clarity respectively. It took a while, but Lexus models became more complex and aggressive, while a handful of Toyotas, most notably the C-HR crossover, took on a deliberate shock value. Here are some of the more intriguing styles and how they came about:

Ford debuted New Edge styling with the Ka and first Focus in 1998; its swoopy curves cut by sharp points made way for Martin Smith's less clearly defined Kinetic Design in the third generation Focus in 2010. Mazda has been one of the most prolific publicizers of design themes, among the most distinctive being Nagare, or flow, under Laurens van den Acker (now Renault design director) in the mid-2000s, and today's Kodo, soul of motion.

BMW's Flame Surfacing was a very influential theme under Chris Bangle from 2001: its complex combination of creases and compound curves in normally featureless panels added reflections, highlights, and allure. Initially criticized, it was soon imitated by other automakers and led directly to the much more complex surfacing employed by Mercedes in particular—though that firm has recently backed away from complexity with its new Sensual Purity style, majoring on form rather than line.

Often, it is those companies who do least to boast about their latest design philosophies that have accomplished the most—either in a profound change of house style or further reinforcement of an already solid brand image.

ANTHONY LO
VICE PRESIDENT OF EXTERIOR DESIGN, RENAULT

Hong Kong-born Anthony Lo joined Renault in 2010 after a career at Lotus, Audi, Mercedes-Benz, and ten years in GM Europe's advanced design studios. Here he talks about how design works at Renault—a company which has renewed its entire range since 2012.

What are the main stages of the product creation process at Renault?
At what we call the kick-off, product planning will come with a brief: what kind of car we are doing, is it a replacement or a new product, who are the customers, and what are the main dimensions. From design we will usually already have some ideas of what this car might look like—a kind of design vision.

How far in advance do you begin developing this vision?
The advanced design phase begins about a year before the kick-off. We'll work on managing proportions, deciding on the character. We test various concepts and ideas, and by kick-off we will have decided on one; we would be clear by this stage and normally have a model, either in clay or white foam to show what kind of car we are talking about.

What happens after the kick-off?
We involve all our designers. Every designer in the whole worldwide group, including our satellite studios in Brazil, India, Romania, China, and Korea, has a chance to work on the project. So we can have a global design competition on all our important programs and concept cars. We get lots of ideas and pick the best five, which we digitize and make scale models of. We then narrow it down to three, which are made into full-size models, and then one is chosen. All this will have happened within about a year of the kick-off. In the last phase, when we go to full size, we work closely with engineering.

Are concept cars developed in the same way?
Yes, but the program is much more condensed because we do not need to worry about feasibility, as it is more or less a design statement. We directly pick the sketches, perhaps three, make the scale models and "go with one" much earlier. The program is about fourteen months from start to finish.

When do you start work on the interior?
It's the same kick-off point, but interiors can take longer so we go for an earlier convergence of themes. While there may still be two themes for exteriors, we're down to a single interior. There are a lot of technical parts and modules, and one area that's relatively new is UX—user experience—design. It's essentially the user interface, everything from the amount of screens we have in the car to the welcome message when you get into the car.

GLOSSARY: DESIGN STUDIO

Advanced Design: Section or group of designers within a studio who work on longer-term ideas and who may also create concept vehicles and undertake early work on programs destined for production.

Alias: Supplier of a wide range of design and presentation software suites employed in the creative industries.

Buck: A prototype interior containing seats, steering wheel, and dashboard structures. Used to try out instrument panel layouts for reach, visibility, and aesthetics.

CAD: Computer-aided design.

Clay: Wax-based clay, which does not dry out, used for quarter- and full-scale models. It can be trimmed, shaped, added to, and smoothed to generate body surfacing features. Final clays can be "dressed" and painted realistically.

Color and Trim: Department within a design operation that deals with finishes, textures, materials, seating, and soft fittings.

Connectivity: Auto industry term for the ability to integrate smartphone functions such as Internet, navigation, music, video, and messaging.

Design Language: In general, the style and character of the surface treatment employed in a vehicle's design. Specifically, each manufacturer's house style or approach to design, often named and used for publicity purposes.

Digital Design: Catch-all description for design using computers rather than paper or physical media.

Downsizing: Trend in the early 21st century toward smaller engines and, among consumers, to move to smaller vehicles as congestion levels increase.

Job One: The first production model off the line for commercial sale.

Kick-off: Term used within Renault for the formal approval and start of a vehicle development program.

Milling: Large computer-driven tool that can machine clay models, including full-sized clays, in three dimensions to highly accurate measurements stored in the digital model.

Model Cycle: Period a particular model stays on the market before being replaced by a fresh model. Most carmakers operate on a cycle of five to seven years, though interim facelifts can prolong a model's sales life.

Photoshop, Illustrator: Semi-professional software programs that can assist in aspects of the digital design process.

Product Planning: A vehicle maker's long-term strategy for vehicle launches, replacements, and other product actions. Generally extends some ten to twelve years, or two model generations, ahead.

Quarter-scale Model: Generally made of clay, ¼-scale models are used to refine alternative proposals at an early stage of the design process.

Rendering: Adding color, texture, reflections, and highlights to two-dimensional sketches or 3D model surfaces.

Scanning: Machine that scans components or models in all three dimensions to build a very accurate digital model of the design. Used frequently in the design process to "save" the design data at each iteration.

Styling Freeze: Point at which all further changes to the vehicle's exterior design are frozen so engineering work can progress unobstructed. Generally two to three years from the on-sale date.

3D Design: Computer-based design where a digital model of the product can be rotated in three dimensions and viewed on screen with different light sources, colors, and special effects

3D Printing: Additive manufacturing where components and parts are built up in metal or other compounds up by adding many minute layers of material. Currently feasible for small prototype parts but too slow for general manufacturing.

Trend Analysis: Commercial subscription services which aim to forecast trends and predict upcoming fashions and retail developments.

UX Design: User experience design. A new field within car design, concentrating on the interface with the consumer—especially electronic systems, display screens, and brand reinforcement messages such as welcome sequences.

Virtual Reality (VR): Electronically created environment allowing interaction between the user and electronic versions of products under evaluation. Can involve full-size holographic displays that allow designers to walk round and inspect "virtual" full-size vehicle proposals. Also used for presentations of designs to senior management.

AND WHAT COMES NEXT?

INTRODUCTION
THE ROAD AHEAD

As we approach the end of the second decade of the 21st century we can look back at the 130 years of car design that have given shape to the automotive culture we hold so dear (many of the finest examples of this heritage have been saluted in this book). But though the astonishing transformations that have taken place in style, engineering, and performance are cause for celebration, on a deep-down level the car is essentially the same thing it has always been—a four-wheeled device designed to take a driver and passengers wherever they want to go. That's why most people love cars.

Yet we must also acknowledge that this is a formula that has, if anything, proved to be too successful for it own good. Everyone wants to be a member of the automotive society; there are more than 1.2 billion vehicles in circulation, pushing the planet to and beyond its limits, choking cities with traffic and causing death and injury on an epidemic scale. Nearly half a century of legislation to improve safety and curb emissions has failed to keep pace with the world's hunger for new cars, and we are now at a tipping point in both how we power our vehicles and how we are permitted to enjoy their privileges will have to change.

These thoughts have crystallised around a handful of future solutions—some have already begun to gain traction—but most are still the stuff of political rhetoric and technical debate among major man-ufacturers and those who represent them. Battery-powered cars, with zero emissions at the point of use, are seen as a response to the issue of urban air quality and are already selling in small numbers in mature markets; hydrogen-powered hybrid vehicles are being tested in California, and semi- or fully automated "driverless" cars are now being promoted as a solution to the issues of safety and the monotony of the regular commute.

Much more tenuous, mainly because it would involve centralized political commitment and signifi-cant infrastructure spending, is the idea of wholly different transport networks to provide eco-friendly mo-bility within megacities. Implicit in any urban mobility plan would be a new type of dedicated city vehicle, a blurring of distinctions between public and private transport, and—most sensitive of all—restrictions on the use of the non-guided, combustion-engined vehicles we depend upon.

There is widespread fear among car enthusiasts that these future scenarios will take the fun out of cars and driving (if indeed the driver is still allowed to take the wheel at all). Some pundits even pre-dict the end of the automotive era as we know it. Yet, on the evidence of Tesla—the youngest and most future-oriented car company—those worries could be unfounded: the 100 percent electric Model S is very fast, very safe, fun to drive, and capable of a high degree of automation. It ticks many of the most import-ant future-scenario boxes and looks attractively contemporary rather than scarily sci-fi, and perhaps its non-confrontational approach is the type of answer that's needed.

AND WHAT COMES NEXT?
CONTINUITY OR CLEAN BREAK?

Like it or not, big changes are approaching. Personal transport will need a fundamental rethink to suit a carbon-neutral future; new types of vehicle will appear and, as we have already seen with Tesla, new automakers and new brand names will emerge as the visible face of the new order. But will the traditional manufacturers adapt and survive, or might they collapse such as Kodak did, when it failed to adapt to the digital photography revolution?

It's a fair bet that when Ford product planners came up with the genius idea for the Mustang in the early 1960s they could not have imagined it would be such a phenomenon, nor that it would still be going strong after more than fifty years. But while the Ford pony car and its sparring partner from Chevrolet, the Camaro, have had their ups and downs over many generations, they are special cars appealing to enthusiast customers to whom the usual rules don't apply. Both are fast, powerful, and indulgent—but will their brand owners be smart enough post-2030 to envisage an e-Camaro or a hydrogen Mustang?

Running on utility rather than adrenalin is another longtime grudge match, spanning a dozen generations since 1966—the Honda Civic versus the Toyota Corolla (now Auris in Europe). The Honda was the more ecological of the two, with the clever 1972 CVCC engine, and though Honda had its first hybrid in 2000, Toyota's parallel solution was much better, and since then Honda has visibly trailed its rival in the eco stakes. But how will these two extend their thinking into the zero-emissions future? Will an electronics outsider sneak in to swindle the most lucrative sales?

The Volkswagen Golf in the mid market has seen an equally fascinating tussle between evolution and revolution. Emerging fresh, slick, well-made, and fun in 1974, it swiftly became the car to beat; now, after eight generations, it is a mature product and VW has its bases covered with almost any powertrain. But will tomorrow's eco Golf still be a recognisable Golf—or will a fresh identity be required for the carbon-free road ahead?

FIRST IN, OR FAST MOVER?

Does it pay to be first with a new technology, or is it better to be a fast follower? The de Havilland Comet was the first commercial jet liner, but it was Boeing who later cashed in big; NSU and Mazda pioneered rotary car engines, but no one followed. Honda had the first hybrid product but Toyota is the market leader; Nissan and Tesla are today's electric innovators—but will they be the ones to reap the long-term rewards?

PREMIUM PRIZE FIGHT

BMW and Mercedes have been locked in a technology race since the 1960s when BMW challenged Mercedes' ascendancy. BMW scored with the first V-12 engine, Mercedes was first with ABS brakes and stability control; BMW pioneered iDrive controls, Mercedes had seven-speed automatics and collision avoidance. Both have plug-in hybrids, but who will be first for full driverless operation?

AND WHAT COMES NEXT?
THE 2020s TAKE SHAPE

KILL WITH STYLE

For a big-volume brand like Volkswagen, the ID concept cars are brilliant marketing. The Golf-sized ID hatch reassures buyers that electric cars can be sleek and attractive, while the ID Buzz and ID Crozz stake out territory for zero-emission family cars and crossovers too.

BMW DOES IT DIFFERENTLY

As the first automaker to launch an in-house electric brand with the i3 in 2013, BMW is well advanced with its autonomous-capable iNext in 2021. This is likely to be radical in its design, underlining the firm's belief that its big technical advances should be made visible with bold exterior style.

DESIGN INTRIGUE

Korea's Hyundai as well as Japanese automakers Honda and Toyota are betting heavily on hydrogen fuel cells. But why do the latter pair's bespoke designs have to be convoluted and unbalanced in the advertising of their advanced engineering?

Established automakers are understandably anxious about the future. One hundred years of what they know best—engine development and chassis dynamics—could be thrown out as zero-emissions regulations begin to bite. In place of the charismatic combustion engine, for so long the defining element of each maker and each model, will come permutations of electric power, each devoid of the individual character that is such a key constituent in every automaker's brand ethos.

Yet with that sudden shift in technology comes a big opportunity for new players. Younger, more agile, and unburdened by heavy old-technology overheads, they could snatch first-mover advantage and secure a profitable foothold in new-era mobility. Tesla, arguably, is the first and best example of the first truly 21st-century automaker, and there are sure to be more—especially from fast-moving China, already the world's largest market for both electric and conventional vehicles.

Several leading traditional automakers are already setting out their stalls to lay claim to (and impose their brand values upon) different patches of future 2020s electric vehicle turf. Volvo, for instance, will turn its Polestar performance sub-brand into a standalone electric car marque, clearly paralleling Tesla in its ambitions. In 2016, Jaguar presented its i-Pace concept for a compact electric crossover: that car will reach the showrooms in 2018, ahead of its key competitors from the German premium brands.

Mercedes' EQ electric car sub-brand will launch a crossover utility in late 2018, most likely followed by a lower-cost hatchback and other derivatives. Rival BMW already has plug-in hybrids at several levels, as well as the i3 hatch and i8 sportscar; GM's Volt and Bolt became staples on the North American market from 2016.

The biggest assault will be that of the VW group, which has already shown sleekly styled electric concepts under the VW, Audi, Porsche, and Bentley nameplates. Volkswagen's fresh-looking ID electric hatchback will appear in 2019, followed by a family of variants; Audi's e-tron crossover is due in late 2018. The same technology will also appear under the hoods of the company's familiar model lines.

AND WHAT COMES NEXT?
NEW POWER, NEW FORMS

HYDROGEN DILEMMA

Long-range hydrogen-fuelled electric cars are harder to package efficiently than battery models. The hydrogen has to be stored at high pressure in rigid cylinders, which could be under a high floor or in the trunk; fuel cell stacks, which generate the current, are becoming smaller.

RECHARGING ON THE MOVE

Renault and other automakers are evaluating systems to allow electric vehicles to recharge without stopping. Special sections of road are fitted with inductive circuits that interact with the moving vehicle to transfer current to the batteries.

SOLAR GAIN

Some electric vehicles already have solar strips on their roof to generate current which then operates a fan to keep the interior cool. Soon, larger areas of solar panels, or perhaps fold-out solar collectors, could generate much greater currents to top up the vehicle's traction battery.

For a designer creating a car shape from scratch, the standard mechanical components of the vehicle are an irritant. They're bulky, hard, and noisy and, because they are fixed in their locations, tend to restrict the designer's freedom in the proportions of the final shape he or she is able to draw around the passenger compartment and chassis.

But with electric cars it could be very different. The largest component, the battery, can be placed anywhere, even split into several parts around the platform. The drive motor is smaller, and could also be replaced by smaller motors on each wheel, and the electronic control unit can again be placed almost anywhere.

These new design freedoms could lead to a wholly new typography of vehicles provided that consumers are prepared to accept these new shapes. The optimum position for the battery in terms of handling is centrally under the floor: this will make most vehicles slightly taller, and designers will have to handle this extra height skillfully, perhaps by specifying larger wheels to disguise the depth of the car's sides. Sports models, however, could have the battery behind the seats to achieve a lower profile.

Without the bulk of a standard engine and transmission in front of the dashboard, long hoods will no longer be required. All that's needed is a defined crush space, so car fronts could become shorter, meaning that the windscreen pillars could move forward, too, allowing a steeper screen slope for better aerodynamics and thus better range from each battery charge.

Overall lengths could shorten as the reduced need for frontal overhang allows a longer wheelbase and more interior space. And with little need for cooling air, the grille could become a thing of the past, though designers would then have to find new ways of projecting brand identity. And that, in a nutshell, is the challenge the next generation of designers will face: successfully transplanting today's carefully-nurtured brand values onto the different vehicle shapes of tomorrow.

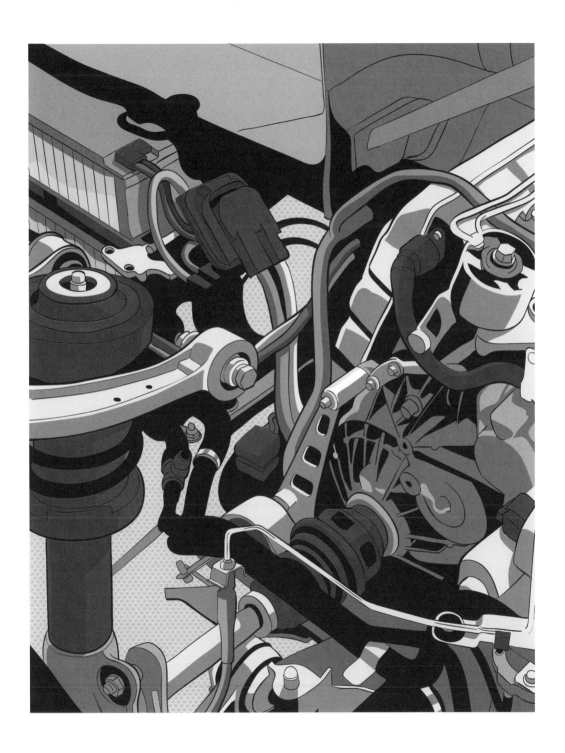

AND WHAT COMES NEXT?
THE URBAN DIMENSION

PARIS AUTOLIB'
Since 2011 Paris has had a public rental scheme, Autolib', for electric cars: users sign in, get their badge, and rent one of 4,000 compact Bluecars from 6,000 locations for €14 an hour. Recently, the Utilib' scheme for small van rental has been added.

GUIDED OR UNGUIDED?
If enough car users simply swapped to electric vehicles once they reached the city limits it would certainly help air quality, but it might not help with traffic congestion and travel times. So it is an open question whether urban mobility vehicles should be automatically guided around the jams or independently steered by the driver.

NEXT-GENERATION TAXIS
Small urban mobility vehicles could be the perfect inner-city taxis. Get in, type in the address of your destination, and enjoy the automated door-to-door ride. But what would all the professional taxi drivers do?

Firstly and most clearly, personal transport—it may no longer be described only as *cars*—will need to be carbon-neutral overall. This means either electric or hydrogen power within cities, where air quality is already at a critical level, and for longer journeys perhaps hydrogen fuel-cell hybrids or even highly efficient combustion-engined hybrids running on synthetic or renewable fuels.

It is in cities where things will need to change most as today's mix of lengthy traffic jams and dirty diesel and petrol engines is toxic to the environment and to people's health. We are already seeing the first generation of small and handy electric vehicles, but these are something of a compromise: tomorrow's megacities, with populations of over ten million, may demand a more focused solution.

One scenario is that we will see a split in functions: bigger and better battery models for out-of-town use and a new generation of dedicated smaller urban mobility vehicles which could be either privately owned or operated by the municipality and rented by the hour or the day, much like the AutoLib' EV fleet in Paris or bike rental stations in dozens of cities.

These vehicles would not need to look like conventional cars as they would be smaller and lighter, influenced by motorcycle and product design as much as conventional auto thinking. Already, design agencies have come up with convincing proposals that also include docking stations and consumer interfaces, but it remains to be seen how much value the established automakers would place on extending their branding onto a city-center mobility service that acted as a halfway house between public transport and the independence and privacy of a personal vehicle.

Some automakers have declared themselves providers of branded mobility services rather than mere car builders. So would it be an attractive prospect to pull off the motorway, park the car, and board a possibly Ford-branded mobility vehicle for the final miles into the city center? Will designers be able to make this an integral part of the brand experience?

AND WHAT COMES NEXT?
THE NEW ERA: SELF-DRIVING VEHICLES

THE FIVE STAGES
OF AUTONOMY

The US-based Society of Automotive Engineers has identified five stages on the route to total autonomous operation. Stage 1 is feet off, with the car doing the accelerating and some braking; Stage 2 allows one hand off, the car taking care of some steering. Stage 3, where we are now with some advanced models, is conditional automation where the car operates itself but the driver may be asked to intervene; Stage 4 provides fully automatic eyes-off travel in specific conditions, but not everywhere. Stage 5, finally, is complete door-to-door automated travel.

WILL DESIGN STILL MATTER?

There are some who argue that with fully automated vehicles and declining interest in the activity of driving, the exterior design and branding of cars will no longer matter. Designers and brand owners naturally disagree.

What if the driver never needed to steer at all? Nor accelerate or brake? The steering wheel, gear change, and pedals could all be eliminated, as could most of the instruments, switches, and sticks. All that would be needed would be a stop-start command and a destination input interface for the navigation.

In terms of design, this changes everything—especially the interior. There need no longer be a designated driver's seat: the command system could instead be a simple hand-held remote, and with the seats turned round to face each other, the cabin would cease to be a technical space dedicated to the task of driving. Instead, it would become a social space, more akin to a lounge or office, and with an ambience in each case of either comfort and luxury or business efficiency.

For designers this would involve a complete rethink of priorities. What will people do if they don't have to drive and the enjoyment of driving is no longer a factor? And with safety having defined the architecture of cabins for so long, how would occupants feel sitting cosily face to face with their companions, as if around a table in a restaurant? Would the free space be liberating, or intimidating?

Designers are working on these issues and more right now, looking for cues from other forms of transport such as train carriages, first-class cabins on airlines, and the interiors of business jets. How would the automotive branding and design cues we know so well be transplanted into such an environment?

Exterior design would face parallel challenges, most notably that of distinguishing autonomously driven vehicles from standard traffic. But would the end result still be thought of as a car? Our ideas around this are sure to evolve, and it will certainly be in the interests of the automotive marque owners to deploy the very best design skills to migrate today's precious brand loyalties to the futuristic vehicles of tomorrow.

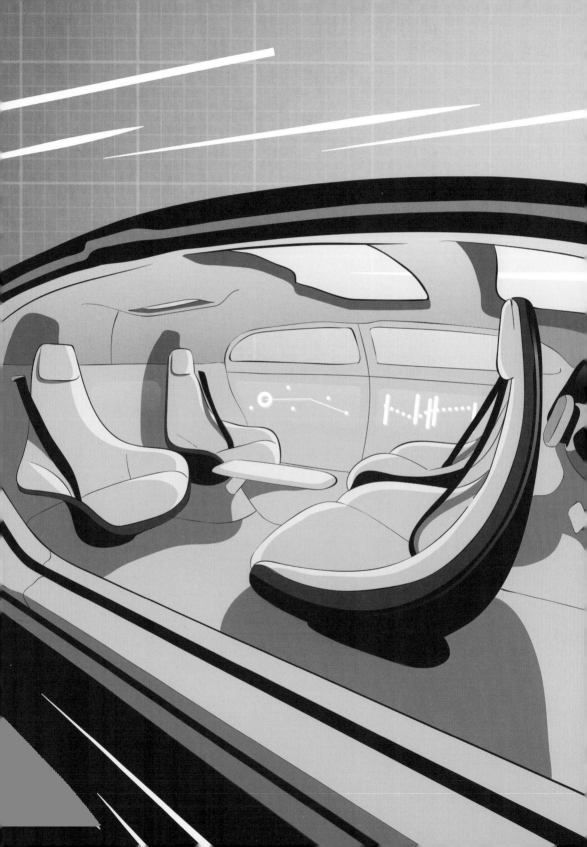

MERCEDES VERSUS BMW:
BATTLE FOR THE FUTURE

The brave new world where vehicles drive themselves is regarded with foreboding and horror by any auto enthusiast who loves cars and loves driving. Nothing could be worse than the prospect of climbing aboard an anonymous digitized plastic pod, typing in a destination, and having nothing to do for the rest of the journey.

With little to go by apart from science fiction stories, there is understandable apprehension about what may eventually prevail. But two recent and contrasting concept cars have been useful in providing visible substance to at least some of the ideas bandied around by the future predictors.

Of the two, from German premium adversaries BMW and Mercedes Benz, it is the latter that may cause more sleepless nights within the auto enthusiast community. Unveiled in 2015, the F015 Luxury in Motion has a large, smooth fuselage with four doors opening to reveal a huge, ice-cool white leather interior housing four rotating chairs in a face-to-face formation. If the driver wants to drive, he or she swivels to the front and operates controls which appear from the dashboard. The windows are not standard glass but are screens on which images are projected; Mercedes describes it as a shared space and private retreat.

BMW's Vision Next100, created to celebrate the company's centennial birthday in 2016, provides what many will feel is a more reassuring foretaste of the future. Despite its sleek supercar profile, it is a four-seater sedan the size of today's 5 Series; all four seats face forward, too, but the interior changes character when the driver shifts from Ease (automated) mode to Boost, for sporty driving. A handlebar-style steering wheel emerges from the dashboard, instrument readings and navigation are projected onto the windshield, and even recommended cornering lines can be displayed. Most dramatic of all is the Next100's copper-colored bodywork: it fully wraps all four wheels and is capable of changing shape as the wheels steer from side to side.

Best of all, BMW has signalled that the iNext, its big push into the future, and directly inspired by the Next100, will appear as a production car in 2021. Perhaps the coming decades won't be so bad after all.

AND WHAT COMES NEXT?
WHAT MAKES A CLASSIC

FULLY QUALIFIED CLASSIC
Consider the Renault Avantime. A high-riding MPV with two giant doors and a chopped-in tail, it launched in 2001 to general bafflement. Two years later, the contract builder assembling it went bust, with barely 8,500 made. Rare, esoteric, and plausibly the first coupé crossover, is it an undiscovered classic?

BRUTISH BENTLEY
Will the future favour the Bentagya, Bentley's first venture into the super-luxury SUV stakes? Big and brutish, it is a spectacular performer but is not blessed with good looks. Will tomorrow's Bentley collectors dote on it or despise it?

WINE THAT AGES WELL
Some designs preserve their good looks over many decades—witness Alfa Romeo's 1954 Giulietta Sprint. But others, even from top-notch marques, don't always age well: the Aston Martin DB7, Ferrari 456 GT, and the 1990s Jaguar XK8 seem less attractive now than in their heydays.

What makes a classic? It is an endlessly debated topic and standard thinking tends to award classic status for value, prestige, rarity, beauty, or originality. So a limited edition Ferrari would score on all counts, a technical pioneer such as a Citroën DS would clearly qualify, as would a Jaguar E-Type—not especially rare, but beautiful—and even surviving examples of big-volume products like the Ford Model T would be included.

All too often, however, there can be a tendency to confuse rarity with obscurity, unworthy products produced in small numbers miraculously gaining classic status. This can have the unintended consequence of rewarding failure rather than success. Some failures are glorious and attractive, like that of the BMW 507 sports car, quite rightly highly prized as only a few hundred were made; others, like the Austin Ambassador, are obscure for good reason, but nevertheless cherished as a historical oddity.

But what about positive historical significance? Engineering landmarks? Early Beetles and Minis are keenly collected, as are Turbos from BMW and Saab. The last of the air-cooled Porsche 911s is now much prized, but the 1997 Toyota Prius, as the first hybrid, is ignored. Will future collectors treasure the Nissan Leaf as the first big-selling electric car? Perhaps it is too soon to say: after all, the wine needs to mature in the bottle before the quality of each vintage becomes clear.

Sentiment plays a big part in determining what is cherished or chucked out. But such emotions are retrospective, and with everyday items—especially those as big as cars, which can't be tucked away at the back of a drawer—it is often hard to see at the time whether they might have a value in the future. Should that rotted-out Ford Taurus be restored? Might that shaky Subaru Impreza be spared its trip to the crusher because it could count as a classic in 2030? And would it rank alongside acknowledged greats such as the Alfa GTV, BMW 3.0 CSL, or Audi quattro? Or even the humble survivors of big-numbers models like the Morris Minor, VW Beetle, and Fiat 500 that we have come to cherish?

AND WHAT COMES NEXT?
COULD COMPUTER-CONTROLLED CARS BECOME CLASSIC?

AN AUTONOMOUS CLASSIC?
Anyone who doubts whether an automated car could one day become a collector's piece should consider computer games. Not the latest, but the earliest ones: once the diet of thrift stores and garage sales, these pioneering consoles and games now fetch high prices.

VINYL: RETURN OF THE OLD
Many audiophiles have long insisted that vinyl records sound better than digital tracks, and a parallel revivalist community is in full swing. Might this be the case with cars in fifty years' time, when auto-pilots are the rule and human drivers can only take the wheel on closed circuits?

BMW iNEXT
At BMW's centenary celebrations in 2016, the CEO promised that the company's big leap forward into the autonomous era would come in 2021 with the launch of its iNext model. Could that model be a safe tip for classic status in 2071?

We have seen in previous chapters that successful car design is defined by a multitude of factors. Proportion and stance, harmony of surfaces, volumes and masses, the quality and appropriateness of detailing; fitness for the intended purpose. Yet these attributes alone are rarely sufficient to guarantee classic status: among the added values most avidly sought out by collectors are technical significance and, for the brave, the feeling that the vehicle is of historical importance and marks a technology landmark of some kind. The first Turbo from BMW would be an example.

Extending that thinking, technical game-changers should be high on any collector's wish-list. BMW's carbon-fibre i3, Audi's aluminium A2, and Tesla's pioneering Model S electric all started something big and could deserve to be tipped as future classics. Now we are on the threshold of a new era, a new order that threatens to render everything we know obsolete. So which will we come to treasure—the last of the old or the first of the new?

Evidence from other sectors where major technology shifts have turned everything upside down is inconclusive. Among rail enthusiasts, steam locomotives are highly prized, but there is little interest in early diesels and electrics. Typewriters became obsolete overnight as computers swept in in the 1980s—but who cares for either? Likewise, in the world of photography, digital had demolished film by the turn of the millennium—so who still remembers Kodak or clings on to a 35mm SLR?

The advent of driverless vehicles will crystallize that first-or-last dilemma. What will people make of the first autonomous car, whatever that turns out to be? Or will we instead celebrate the last car capable of being steered by a human driver? And then, a generation later, will the same thing apply in 2050 to that pioneering 2020s autonomous car? Or, perhaps more to the point, will the proud owner still be allowed to ride in something so old? Only time will tell.

GLOSSARY: FUTURE CAR

Air Quality: A serious health issue in major cities in the early 21st century, some of the worst effects being attributable to nitrogen oxides and soot particle emissions from combustion-engined cars, taxis, buses, and trucks.

Aluminium Space Frame: Type of car construction where the structure is made up of a framework of small aluminium load-bearing members, covered by an external skin. Lighter than most conventional structures.

Autonomous Vehicle: Vehicle capable of undertaking some or all driving tasks using its own systems to relieve the load on the human driver. The US-based Society of Automotive Engineers has identified five stages on the route to total autonomous operation. Stage 1 is feet off, with the car doing the accelerating and some braking; Stage 2 allows one hand off, the car taking care of some steering. Stage 3, where we are now with some advanced models, is conditional automation where the car operates itself but the driver may be asked to intervene; Stage 4 provides fully automatic eyes-off travel in specific conditions, but not everywhere. Stage 5, finally, is complete door-to-door automated travel.

Carbon Neutral: A device or a process where emissions of carbon are balanced by carbon absorption or offsets. Some second-generation biofuels come under this heading.

Carbon Fiber: Light and very strong structural material used principally in racing and supersports cars. Employed in BMW i3 and i8 electrified vehicles to offset extra weight of batteries.

Connectivity: Auto industry term for the ability to integrate smartphone functions such as internet, navigation, music, video, and messaging.

Driverless Vehicle: See **autonomous vehicle**

EV, BEV: Electric vehicle relying on battery power alone, with no auxiliary engine or power source.

Greenhouse Gas (GHG) emissions: The principal GHG emission from motor vehicles is carbon dioxide (CO_2), the leading contributor to global climate change. CO_2 emissions are directly proportional to the amount of fuel burned, but other emissions such as methane (CH_4) and nitrous oxide (N_2O) also have greenhouse effects.

Grid Balancing: The use of plugged-in electric vehicles to absorb, store, and sometimes even return surplus electric power to national or local electricity grids. Can effectively increase the capacity of electricity networks and help smooth out peaks and troughs in supply and demand.

Hydrogen Fuel Cell: A propulsion system that mixes hydrogen fuel and oxygen from the atmosphere to produce electric current and power a traction motor. The fuel cell has no moving parts and its only emission is water.

Megacity: Very large city with a population exceeding 10 million, typically extending over a large area and often the result of large-scale migration from rural areas. Tokyo and Jakarta are currently the largest, at 38 and 31 million inhabitants, respectively.

Modal Shift: Process of switching from one form of transport to another, such as parking a car to catch a train, or swapping a long-range car for a dedicated urban vehicle at a suburban interchange point.

Networking: The electronic linking of different types of transport users with the transport infrastructure to coordinate traffic flows, maximize system capacity, and minimize emissions and energy use.

PHEV: Plug-in hybrid vehicle. A hybrid with a more powerful electric motor and a bigger battery to allow it to operate under pure electric power for some distance and to be recharged from grid electricity.

Pollutant Emissions: Emissions from vehicle exhausts of toxic compounds including soot/carbon particulates, nitrogen oxides, and volatile compounds. Responsible for deterioration in urban and regional air quality as well as for upper atmosphere effects such as ozone layer depletion.

Urban Mobility Vehicle: Dedicated electric vehicle suitable for use in future megacities as an alternative to conventional vehicles. Seen as a possible solution to urban air quality problems.

Zero Emissions: Vehicle or system which gives off only harmless emissions, such as water, at its point of use, though this may have the effect of displacing those harmful emissions to a power plant elsewhere. Battery and hydrogen vehicles are zero emission at their point of use.

KEY DESIGNER TIMELINES

HARLEY EARL, CHAPTER 1

1893: Born in LA

1918: Director of custom bodyshop

1920: Meets Alfred P Sloan, president of General Motors

1927: Launches GM's Art and Color section, designs Cadillac La Salle

1933: Cadillac V-16 Aero-Dynamic Coupé

1938: Buick Y-Job, the first concept car

1940s: Vice president of GM

1947: Buick Roadmaster

1948: Cadillac introduces tail fins

1949: Buick, Cadillac, Oldsmobile hardtops

1951: GM Buick LeSabre concept car

1953: Chevrolet Corvette; first Motorama travelling roadshows

1954: Pontiac Firebird XP-21 gas turbine research car

1958: Firebird II concept

1959: Cadillac Eldorado

PININFARINA, CHAPTER 2

1930: Battista 'Pinin' Farina sets up Carrozzeria Pininfarina

1931: Lancia Dilambda

1935: Alfa Romeo 6C Pescara coupé

1947: Cisitalia 202 GT coupé

1952: Alfa Romeo 1900 Sprint Coupé

1954: Lancia Aurelia B24 S

1955: Alfa Romeo Giulietta Spider, Peugeot 403

1958: Austin A40 Farina, Ferrari 250GT

1961: Company renamed Pininfarina as Battista hands over to son Sergio

1966: Battista 'Pinin' Farina dies; Alfa Spider 1600 Duetto ('Graduate')

1968: BMC 1800 'aerodinamica' concept

1969: Peugeot 504 Coupé and Cabriolet

1971: Ferrari Berlinetta Boxer

1978: Jaguar XJ Spider design study

1979: CNR 'banana car' aerodynamic design study

1983: Peugeot 205

1986: Cadillac Allante, design and manufacture

1987: Ferrari F40 marks 40th anniversary of Ferrari

1992: Ferrari 456 GT

1995: Alfa Romeo Spider and GTV

1997: Peugeot 406 Coupé

1999: Ferrari 360; Metrocubo city car study

2004: Ferrari F430

2005: Alfa Romeo Brera; Maserati Birdcage concept

2009: Ferrari 458 Italia; Ferrari California

2017: Ferrari 812 Superfast

CITROEN, CHAPTER 3

1878: André Citroen born in Paris

1919: founds Automobiles André Citroen, employs Ford's manufacturing techniques, launches Type A

1924: Citroen B10 is the first European car with all-steel bodywork

1926: One-third of all vehicles on French roads are Citroens

1932: Rosalie series; Citroen is the fourth-biggest automaker in the world

1934: Advanced Type 7A Traction Avant causes a major stir in auto industry

1934: Insolvent; taken over by tire maker Michelin.

1935: André Citroen dies of cancer

1936: Company begins work on minimalist small car

1938: First tests of 2CV

1948: 2CV

1955: DS 19 causes sensation at Paris show

1961: Mid-market Ami 6

1965: Acquires struggling Panhard company

1968: Takes control of Maserati

1970: SM luxury coupé; compact GS

1974: Rescued by rival Peugeot; CX replaces DS

1978: Visa

1982: Mid-sized BX, with polarizing design by Gandini

1989: New flagship XM, with Hydractive suspension

1990: Production of 2CV ends after 5.1 million units

1998: Xsara Picasso is Citroen's answer to Renault Scenic

1999: C6 Lignage concept car re-establishes
reputation as innovator

2001: C3 hatchback

2005: C1 city car, joint venture with Peugeot and Toyota

2007: Funky C4 Cactus concept

2009: 90th birthday, revives the DS name
for upmarket supermini

2010: DS becomes separate premium brand

2014: C4 Cactus

2017: Peugeot Citroen acquires Opel-Vauxhall

GIORGIETTO GIUGIARO, CHAPTER 4

1938: Born in northern Italy

1955: Joins Fiat as apprentice designer

1959: Joins Bertone, designs Alfa Romeo 2600 Coupé,
Giulia GT, BMW 3200CS

1965: Joins Ghia, designs Iso Grifo and Rivolta,
Simca 1200S Coupé, Maserati Ghibli

1968: Founds Italdesign, designs Bizzarini Manta,
Maserati Bora, Alfa Romeo Alfasud

1971: Begins work on Volkswagen Passat, Scirocco, Golf

1974: Alfetta GT, Alfasud Sprint coupé; VW Golf

1975: Lotus Esprit introduces 'folded paper'
design language

1978: Lancia Megagamma concept; BMW M1, Audi 80

1979: Lancia Delta

1980: Fiat Panda low-cost car; stays in production 23 years

1981: DeLorean DMC12, Isuzu Piazza

1983: Fiat Uno introduces tall proportions to
supermini sector

1984: Saab 9000, Lancia Thema, Isuzu Gemini/Chevrolet
Spectrum, SEAT Ibiza

1993: Fiat Punto

1998: Maserati 3200GT

1999: Voted Car Designer of the Century

2002: Alfa Romeo Brera, on sale 2005

2003: Lamborghini Gallardo

2004: Fiat Croma, explores crossover-like architecture

2005: Fiat Grande Punto

2007: Fiat Seidici/Suzuki SX4

2010: Volkswagen group buys 90% stake in Italdesign

2015: Giugiaro sells remaining stake and resigns

REFERENCES AND SOURCES

Bayley, Stephen, and Conran, Terence: *The A-Z of Design*,
Conran Octopus, London, 2007

Denison, Edward (ed): *30-Second Architecture*, Ivy Press,
Lewes, UK, 2013

Design Museum London and Nahum, Andrew: *Fifty Cars that
Changed the World*, Octopus Books, London, 2009

Felicioli, Riccardo P: *Sergio Pininfarina*, *Studi & Ricerche*,
Automobilia, Milan, 1998

Felicioli, Riccardo P: *Alfa Romeo, la Bellesa Necessaria*,
Automobilia, Milan, 1994

Lewin, Tony: *A-Z of 21st Century Cars*, Merrell, London, 2011

Lewin, Tony: *The BMW Century*, Motorbooks, Minneapolis,
2016

Lewin, Tony (ed) and Newbury, Stephen: *The Car Design
Yearbook*, vols 1-8, Merrell, London, 2002-2009

Lewin, Tony and Borroff, Ryan: *How to Design Cars Like a
Pro, 1st and 2nd editions*, Motorbooks, Minneapolis, 2003
& 2010

Mason, George: Patrick le Quément, *Renault Design*, Automobilia, Milan, 2000

Polster, Bernd, et al: *The A-Z of Modern Design*. Merrell,
London, 2009

Sparke, Penny: *A Century of Car Design*, Mitchell Beazley,
London, 2002

Strasser, Josef: *50 Bauhaus Icons You Should Know*, Prestel,
Munich, 2009

INDEX